This book is due for return not later than the last
date stamped below, unless recalled sooner.

Biomimetic Nanoceramics in Clinical Use
From Materials to Applications

RSC Nanoscience & Nanotechnology

Series Editors

Professor Paul O'Brien, *University of Manchester, UK*
Professor Sir Harry Kroto FRS, *University of Sussex, UK*
Professor Harold Craighead, *Cornell University, USA*

This series will cover the wide ranging areas of Nanoscience and Nanotechnology. In particular, the series will provide a comprehensive source of information on research associated with nanostructured materials and miniaturised lab on a chip technologies.

Topics covered will include the characterisation, performance and properties of materials and technologies associated with miniaturised lab on a chip systems. The books will also focus on potential applications and future developments of the materials and devices discussed.

Ideal as an accessible reference and guide to investigations at the interface of chemistry with subjects such as materials science, engineering, biology, physics and electronics for professionals and researchers in academia and industry.

Titles in the Series:

Atom Resolved Surface Reactions: Nanocatalysis
PR Davies and MW Roberts, *School of Chemistry, Cardiff University, Cardiff, UK*

Biomimetic Nanoceramics in Clinical Use: From Materials to Applications
María Vallet-Regí and Daniel Arcos, *Department of Inorganic and Bioinorganic Chemistry, Complutense University of Madrid, Madrid, Spain*

Nanocharacterisation
Edited by AI Kirkland and JL Hutchison, *Department of Materials, Oxford University, Oxford, UK*

Nanotubes and Nanowires
CNR Rao FRS and A Govindaraj, *Jawaharlal Nehru Centre for Advanced Scientific Research, Bangalore, India*

Visit our website at www.rsc.org/nanoscience

For further information please contact:
Sales and Customer Care, Royal Society of Chemistry, Thomas Graham House, Science Park, Milton Road, Cambridge, CB4 0WF, UK
Telephone: +44 (0)1223 432360, Fax: +44 (0)1223 426017, Email: sales@rsc.org

Biomimetic Nanoceramics in Clinical Use
From Materials to Applications

María Vallet-Regí and Daniel Arcos
*Department of Inorganic and Bioinorganic Chemistry,
Complutense University of Madrid, Madrid, Spain*

RSCPublishing

ISBN: 978-0-85404-142-8

A catalogue record for this book is available from the British Library

Published by The Royal Society of Chemistry,
Thomas Graham House, Science Park, Milton Road,
Cambridge CB4 0WF, UK

Registered Charity Number 207890

For further information see our website at www.rsc.org

Preface

The research on nanoceramics for biomedical applications responds to the challenge of developing fully biocompatible implants, which exhibit biological responses at the nanometric scale in the same way that biogenic materials do. Any current man-made implant is not fully biocompatible and will always set off a foreign body reaction involving inflammatory response, fibrous encapsulation, *etc.* For this reason, great efforts have been made in developing new synthetic strategies that allow tailoring implant surfaces at the nanometric scale. The final aim is always to optimise the interaction at the tissue/implant interface at the nanoscale level, thus improving the life quality of the patients with enhanced results and shorter rehabilitation periods.

The four chapters that constitute this book can be read as a whole or independently of each other. In fact, the authors' purpose has been to write a book useful for students of biomaterials (by developing some basic concepts of biomimetic nanoceramics), but also as a reference book for those specialists interested in specific topics of this field. At the beginning of each chapter, the introduction provides insight on the corresponding developed topic. In some cases, the different introductions deal with some common topics. However, even at the risk of being reiterative, we have decided to include some fundamental concepts in two or more chapters, thus allowing the comprehension of each one independently.

Chapter 1 deals with the description of biological hard tissues in vertebrates, from the point of view of mineralization processes. For this aim, the concepts of hard-tissue mineralisation are applied to explain how Nature works. This chapter finally provides an overview about the artificial alternatives suitable to be used for mimicking Nature.

In Chapter 2 we introduce general considerations of solids reactivity, which allow tailoring strategies aimed at obtaining apatites in the laboratory. These strategies must be modified and adapted in such a way that artificial carbonated

RSC Nanoscience & Nanotechnology
Biomimetic Nanoceramics in Clinical Use: From Materials to Applications
By María Vallet-Regí and Daniel Arcos

calcium-deficient nanoapatites can be obtained resembling as much as possible the biological apatites. For this purpose, a review on the synthesis methods applied for apatite obtention are collected in the bibliography.

In Chapter 3 we have focused on the specific topic of hard-tissue-related biomimetism. To reach this goal, we have dealt with nanoceramics obtained as a consequence of biomimetic processes. The reader will find information about the main topics related with the most important bioactive materials and the biomimetic apatites growth onto them. Concepts and valuable information about the most widely used biomimetic solutions and biomimetism evaluation methods are also included.

Finally, Chapter 4 reviews the current and potential clinical applications of apatite-like biomimetic nanoceramics, intended as biomaterials for hard-tissue repair, therapy and diagnosis.

The authors wish to thank RSC for the opportunity provided to write this book, as well as their comprehensive technical support. Likewise, we want to express our greatest thanks to Dr. Fernando Conde, Pilar Cabañas and José Manuel Moreno for their assistance during the elaboration of this manuscript. We are also thankful to Dr. M. Colilla, Dr. M. Manzano, Dr. B. Gónzalez and Dr. A.J. Salinas for their valuable suggestions and scientific discussions. Finally, we would like to express our deepest gratitude to all our coworkers and colleagues that have contributed over the years with their effort and thinking to these studies.

María Vallet-Regí
Daniel Arcos

Contents

RSC Nanoscience & Nanotechnology
Biomimetic Nanoceramics in Clinical Use: From Materials to Applications
By María Vallet-Regí and Daniel Arcos
© María Vallet-Regí and Daniel Arcos, 2008

Abbreviations

ACP   Amorphous Calcium Phosphate
ALP   Alkaline Phosphatase
BCP   Biphasic Calcium Phosphate
BG   Bioactive Glass
BSG   Bioactive Star Gel
CaP   Calcium Phosphate
CDHA   Calcium-Deficient Hydroxyapatite
CHA   Carbonate Hydroxyapatite
CTAB   Cetyl Trimethyl Ammonium Bromide
CVD   Chemical Vapour Deposition
ECM   Extracellular Matrix
ED   Electron Diffraction
EDS   Energy Dispersive X-ray Spectroscopy
EISA   Evaporation-Induced Self-Assembly
FTIR   Fourier Transform Infrared (spectroscopy)
HA   Hydroxyapatite
HRTEM   High-Resolution Transmission Electron Microscopy
MBG   Mesoporous Bioactive Glass
OCP   Octacalcium Phosphate
OHA   Oxyhydroxyapatite
PCL   Poly(ε-caprolactone)
PDMS   Poly(dimethylsiloxane)
PEG   Poly(ethylene glycol)
PLLA   Poly(l-lactic acid)
PMMA   Poly(methyl methacrylate)
PVAL   Poly(vinyl alcohol)
QD   Quantum Dot
SA   Serum Albumin
SBF   Simulated Body Fluid
SEM   Scanning Electron Microscopy
SiHA   Silicon-Substituted Hydroxyapatite
TCP   Tricalcium Phosphate
TEM   Transmission Electron Microscopy
TEOS   Tetraethylorthosilicate
TTCP   Tetracalcium Phosphate
XRD   X-ray Diffraction

CHAPTER 1

Biological Apatites in Bone and Teeth

1.1 Hard-Tissue Biomineralisation: How Nature Works

The bones and teeth of all vertebrates are natural composite materials (Figure 1.1), where one of the components is an inorganic nanocrystalline solid with apatite structure and the chemical composition of a carbonated, basic calcium phosphate, hence it can be termed a *carbonate-hydroxy-apatite*. It amounts to 65% of the total bone mass, with the remaining mass formed by organic matter and water.[1]

Most of the biominerals are inorganic/organic composite materials.[2] In this sense, the bones of vertebrates are also formed by the combination of an inorganic calcium phase – *carbonate-hydroxyl-apatite* – and an organic matrix.[3] The benefits that the inorganic part brings to this combination are *toughness* and the ability to *withstand pressure*.

On the other hand, the organic matrix formed by collagen fibres, glycoproteins and mucopolysaccharides, provides *elasticity* and *resistance to stress*, *bending* and *fracture*. Such symbiosis of two very different compounds, with markedly different properties, confers to the final product, *i.e.* the biomineral, some properties that would be unattainable for each of its individual components *per se*. This is a fine example in Nature of the advantages that a composite material can exhibit, in order to reach new properties with added value. In fact due to this evidence, a large portion of the modern materials science field is currently focused on the development of composite materials.

1.1.1 Bone Formation

The bone exhibits some physical and mechanical properties that are rather unusual. It is able to bear heavy loads, to withstand large forces and to flex

RSC Nanoscience & Nanotechnology
Biomimetic Nanoceramics in Clinical Use: From Materials to Applications
By María Vallet-Regí and Daniel Arcos

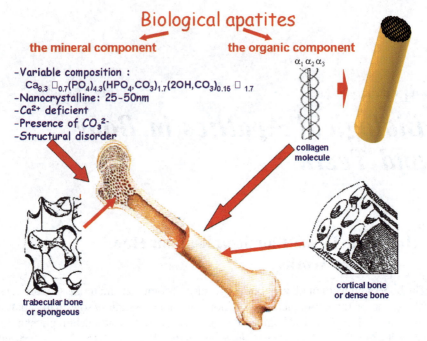

Figure 1.1 Inorganic–organic composite nature of both trabecular and cortical bone.

without fracture within certain limits. Besides, the bone also acts as an ion buffer both for cations and anions. From the material point of view, the bone could be simplified as a three-phase material formed by *organic fibres*, an *inorganic nanocrystalline phase*, and a *bone matrix*. Its unique physical and mechanical properties are the direct consequence of intrinsic atomic and molecular interactions within this very particular natural composite material.

Bone is not uniformly dense. It has a hierarchical structure. Due to its true organic-inorganic composite nature, it is able to adopt different structural arrangements with singular architectures, determined by the properties required from it depending on its specific location in the skeleton. Generally speaking, most bones exhibit a relatively dense outer layer, known as *cortical* or *compact* bone, which surrounds a less dense and porous, termed *trabecular* or *spongy* bone, which is in turn filled with a jelly tissue: the *bone marrow*.[4] This complex tissue is the body deposit of nondifferentiated trunk cells, precursors of most repairing and regenerating cells produced after formation of the embryonic subject.[5,6] The bone fulfils critical functions in terms of a *structural* material and an ion *reservoir*. Both functions strongly depend on the size, shape, chemical composition and crystalline structure of the mineral phase, and also on the mineral distribution within the organic matrix.

The main constituents of bone are: *water*; a mineral phase, *calcium phosphate* in the form of carbonated apatite with low crystallinity and nanometric dimensions, which accounts for roughly two thirds of the bone's dry weight; and an organic fraction, formed of *several proteins*, among which type-I collagen is the main component, which represents approximately the remaining one third of bone dry weight. The other intervening proteins, such as proteoglicans and glycoproteins, total more than two hundred different proteins, known as noncollagen proteins; their total contribution to the organic constituent, however, falls below 10% of the said organic fraction. These bone constituents are hierarchically arranged with, at least, five levels of organisation. At the molecular level, the polarised triple helix of tropocollagen molecules are grouped in microfibres, with small cavities between their edges, where small apatite crystals – approximately 5 nm × 30 nm sized – nucleate and grow. These microfibres unite to form larger fibres that constitute the microscopic units of bone tissue. Then, these fibres are arranged according to different structural distributions to form the full bone.[7]

It was traditionally believed that the inorganic phase was mainly amorphous calcium phosphate that, in the ageing process, evolved towards nanocrystalline hydroxyapatite. Results of solid-state ^{31}P NMR spectroscopy, however, showed that the amorphous phase is never present in large amounts during the bone development process.[6] Besides, this technique did detect acid phosphate groups. Phosphate functions correspond to proteins with *O*-phosphoserine and *O*-phosphotreonine groups, which are probably used to link the inorganic mineral component and the organic matrix. Phosphoproteins are arranged in the collagen fibres so that Ca^{2+} can be bonded at regular intervals, in agreement with the inorganic crystal structure, hence providing a repeating condition that leads to an ordered sequence of the same unit, *i.e.* the crystallinity of the inorganic phase. The cells responsible for most of the assembling process are termed *osteoblasts*. When the main assembling process is completed, the osteoblasts keep differentiating in order to form *osteocytes*, which are responsible for the bone maintenance process. The controlled nucleation and growth of the mineral take place at the microscopic voids formed in the collagen matrix. The type-I collagen molecules, segregated by the osteoblasts, are grouped in microfibres with a specific tertiary structure, exhibiting a periodicity of 67 nm and 40 nm cavities or orifices between the edges of the molecules.[7] These orifices constitute microscopic environments with free Ca^{2+} and PO_4^{3-} ions, as well as groups of side chains eligible for bonding, with a molecular periodicity that allows the nucleation of the mineral phase in a heterogeneous fashion. Ca^{2+} ions deposited and stored in the skeleton are constantly renewed with dissolved calcium ions. The bone growth process can only be produced under a relative excess of Ca^{2+} and its corresponding anions, such as phosphates and carbonates, at the bone matrix. This situation is achieved due to the action of efficient ATP-powered ionic pumps, such as Ca^{2+} ATPases for active transportation of calcium.[8–10] In terms of physiology, carbonate and phosphate are present in the form of HCO_3^-, HPO_4^{2-} and $H_2PO_4^-$ anions. When incorporated to the bone, the released protons can move throughout the bone tissue and leave the nucleation and mineralisation area. The nucleation of thin, platelet-shaped apatite crystals, takes place at the bone within discrete spaces inside the collagen fibres, hence restricting a potential primary growth of

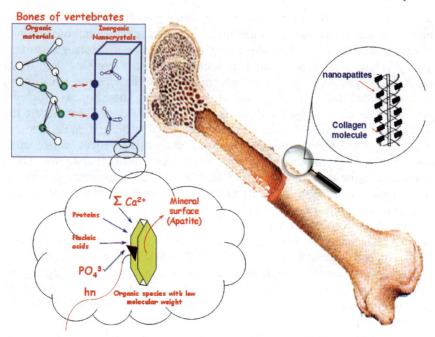

Figure 1.2   Interaction between biological nanoapatites and organic fraction of bone
at the molecular scale. At the bottom of the scheme: formation of
nanoapatite crystallites with the factors and biological moieties present in
the process. A magnified scheme of the apatite crystallites location into
collagen fibres is also displayed.

these mineral crystals, and imposing their discrete and discontinuous quality
(Figure 1.2).

Calcium phosphate nanocrystals in bone, formed at the mentioned spaces left
between the collagen fibres, exhibit the particular feature of being mono-
dispersed and nanometre-sized platelets of *carbonate-hydroxyl-apatite*. There is
no other mineral phase present, and the crystallographic axis *c* of these crystals
is arranged parallel to the collagen fibres and to the largest dimension of the
platelet. In the mineral world, the thermodynamically stable form of calcium
phosphate under standard conditions is the hydroxyapatite (HA).[11] Generally
speaking, this phase grows in needle-like forms, with the *c*-axis parallel to the
needle axis. Figure 1.3 shows the crystalline structure of hydroxyapatite,
$Ca_{10}(PO_4)_6(OH)_2$, which belongs to the hexagonal system, space group $P6_3/m$
and lattice parameters $a = 9.423$ Å and $c = 6.875$ Å.

Besides the main ions Ca^{2+}, PO_4^{3-} and OH^-, the composition of biological
apatites always includes CO_3^{2-} at approximately 4.5%, and also a series of
minority ions, usually including Mg^{2+}, Na^+, K^+, Cl^-, F^-.[12] These substitutions
modify the lattice parameters of the structure as a consequence of the different
size of the substituting ions, as depicted in Figure 1.3. This is an important
difference between minerals grown in an inorganic or biological environment.

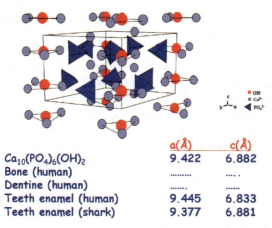

	a(Å)	c(Å)
$Ca_{10}(PO_4)_6(OH)_2$	9.422	6.882
Bone (human)	....	.....
Dentine (human)	....	....
Teeth enamel (human)	9.445	6.833
Teeth enamel (shark)	9.377	6.881

Figure 1.3 Crystalline structure and unit cell parameters for different biological hydroxyapatites.

The continuous formation of bone tissue is performed at a peripheral region, formed by an external crust and an internal layer with connective tissue and *osteoblast* cells. These osteoblasts are phosphate-rich and exude a jelly-like substance, the osteoid. Due to the gradual deposit of inorganic material, this osteoid becomes stiffer and the osteoblasts are finally confined and transformed in bone cells, the *osteocytes*. The bone-transformation mechanism, and the ability to avoid an excessive bone growth, are both catered for by certain degradation processes that are performed simultaneously to the bone formation. The osteoclasts, which are giant multinucleated cells, are able to catabolyse the bone purportedly using citrates as chelating agent. The control of the osteoclast activity is verified through the action of the parathyroid hormone, a driver for demineralisation, and its antagonist, tireocalcitonin.

The collagen distribution with the orifices previously described is *necessary* for the controlled nucleation and growth of the mineral, but it might not *suffice*. There are conceptual postulations of various additional organic components, such as the phosphoproteins, as an integral part of the nucleation core and hence directly involved in the nucleation mechanism. Several immuno-cyto-chemical studies of bone, using techniques such as optical microscopy and high-resolution electron microscopy, have clearly shown that the phosphoproteins are restricted or, at least, largely concentrated at the initial mineralisation location, intimately related to the collagen fibres. It seems that the phosphoproteins are enzymatically phosphored previously to the mineralisation.[13]

The crystallisation of the complex and hardly soluble apatite structures evolves favourably through the kinetically controlled formation of metastable intermediate products. Under *in vitro* conditions, amorphous calcium phosphate is transformed into octacalcium phosphate (OCP) that, in turn, evolves to carbonate hydroxyapatite; at lower pH values, the intermediate phase seems to be dehydrated dicalcium phosphate (DCPD).[14,15]

The mechanisms of bone formation are highly regulated processes,[7] which seem to verify the following statements:

– Mineralisation is restricted to those specific locations where crystals are constrained in size by a compartmental strategy.
– The mineral formed exhibits specific chemical composition, crystalline structure, crystallographic orientation and shape. The chemical phase obtained is controlled during the stages of bone formation. In vertebrates, said chemical phase is a hydroxyl-carbonate-apatite, even though the thermodynamically stable form of calcium phosphate in the world of minerals, under standard conditions, is hydroxyapatite.
– Since the mineral deposits onto a biodegradable organic support, complex macroscopic forms are generated with pores and cavities. The assembling and remodelling of the structure are achieved by cell activity, which builds or erodes the structure layer by layer.

Without a careful integration of the whole process, bone formation would be an impossible task. The slightest planning mistake by the body, for instance in its genetic coding or cell messengers, is enough to provoke building errors that would weaken the osseous structure.

The hard tissues in vertebrates are bones and teeth. The differences between them reside in the amounts and types of organic phases present, the water content, the size and shape of the inorganic phase nanocrystals and the concentration of minor elements present in the inorganic phase, such as CO_3^{2-}, Mg^{2+}, Na^+, *etc.*[12] The definitive set of teeth in higher-order vertebrates has an outer shell of dental enamel that, in an adult subject, *does not contain* any living cells.[16] Up to 90% of said enamel can be inorganic material, mainly *carbonate-hydroxyl-apatite*. Enamel is the material that undergoes more changes during the tooth development process. At the initial stage, it is deposited with a mineral content of only 10–20%, with the remaining 80–90% of proteins and special matrix fluids. In the subsequent development stages, the organic components of the enamel are almost fully replaced by inorganic material. The special features of dental enamel when compared with bone material are its much larger crystal domains, with prismatic shapes and strongly oriented, made of *carbonate-hydroxyl-apatite* (Figure 1.4). There is no biological material that could be compared to enamel in terms of hardness and long life. However, it cannot be regenerated.

The bones, the body-supporting scaffold, can exhibit different types of integration between organic and inorganic materials, leading to significant variations in their mechanic properties. The ratio of both components reflects the compromise between toughness (high inorganic content) and resiliency or fracture strength (low inorganic content). All attempts to synthesise bone replacement materials for clinical applications featuring physiological tolerance, biocompatibility and long-term stability have, up to now, had only relative success; which shows the superiority and complexity of the natural structure where, for instance, a human femur can withstand loads of up to 1650 kg.[17]

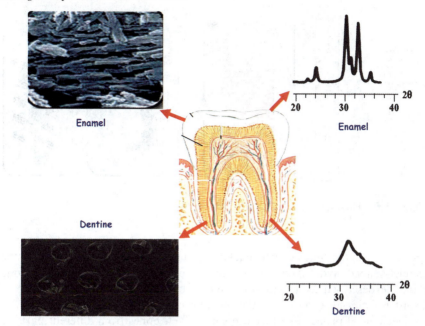

Figure 1.4 Different apatite crystallinity degrees in teeth. Enamel (top) is formed by well-crystallised apatite, whereas dentine (bottom) contains nanocrystal-line apatite within a channelled protein structure.

The bones of vertebrates, as opposed to the shells of molluscs, can be considered as *"living biominerals"* since there are cells inside them under permanent activity. It also constitutes a storage and hauling mechanism for two essential elements, phosphorus and calcium, which are mainly stored in the bones. Most of what has been described up to this point, regarding the nature of bone tissue, could be summed up by stating that *the bone is a highly structured porous matrix, made of nanocrystalline and nonstoichiometric apatite, calcium deficient and carbonated, intertwined with collagen fibres and blood vessels.*

Bone functions are controlled by a series of hormones and bone-growth factors. Figure 1.5 attempts to depict these phenomena in a projection from our macroscale point of view, to the "invisible" nanoscale.

Bone's rigidity, resistance and toughness are directly related to its mineral content.[18] Although resistance and rigidity increase linearly with the mineral content, toughness does not exhibit the same trend, hence there is an optimum mineral concentration that leads to a maximum in bone toughness. This tendency is clearly the reason why the bone exhibits a restricted amount of mineral within the organic matrix. But there are other issues affecting the mechanical properties of bone, derived from the microstructural arrangement of its components. In this sense, the three main components of bone exhibit radically different properties. From this point of view, the biomineral is clearly a *composite*.[19] The organic scaffold exhibits a fibrous structure with three levels: the individual triple helix molecules, the small fibrils, and its fibre-forming

Figure 1.5 Hierarchical organisation of bone tissue.

aggregates. These fibres can be packed in many different ways; they host the platelet-shaped hydroxyl-carbonate-apatite crystals. In this sense, the bone could be described as a composite reinforced with platelets, but the order–disorder balance determines the microstructure and, as a consequence, the mechanical properties of each bone. In fact, bones from different parts of the body show different arrangements, depending on their specific purpose.

Bone crystals are extremely small, with an average length of 50 nm (in the 20–150 nm range), 25 nm in average width (10–80 nm range) and thickness of just 2–5 nm. As a remarkable consequence, a large part of each crystal is surface; hence their ability to interact with the environment is outstanding.

Apatite phase contains between 4 and 8% by weight of carbonate, properly described as *dahllite*. Mineral composition varies with age and it is always calcium deficient, with phosphate and carbonate ions in the crystal lattice. The formula $Ca_{8.3}(PO_4)_{4.3}(CO)_{3x}(HPO_4)_y(OH)_{0.3}$ represents the average composition of bone, where y decreases and x increases with age, while the sum $x + y$ remains constant and equal to 1.7.[12] Mineral crystals grow under a specific orientation, with the c-axes of the crystals approximately parallel to the long axes of the collagen fibres where they are deposited. Electron microscopy techniques were used to obtain this information.[20]

The bones are characterised by their composition, crystalline structure, morphology, particle size and orientation. The apatite structure hosts carbonate in two positions: the OH^- sublattice producing so-called type A carbonate apatites or the $[PO_4]^{3-}$ sublattice (type B apatites) (Figure 1.6).

The small apatite crystal size is a very important factor related to the solubility of biological apatites when compared with mineral apatites. Small dimensions and low crystallinity are two distinct features of biological apatites that, combined with their nonstoichiometric composition, inner crystalline disorder and presence of carbonate ions in the crystal lattice, allow their special behaviour to be explained.

Apatite structure allows for wide compositional variations, with the ability to accept many different ions in its three sublattices (Figure 1.7).

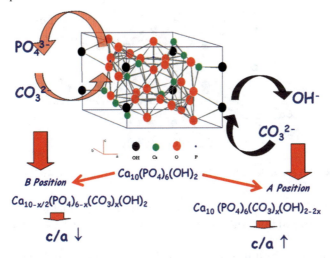

Figure 1.6 Crystalline structure and likely ionic substitutions in carbonate apatites.

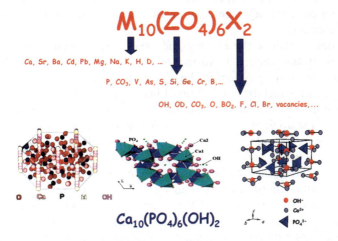

Figure 1.7 Compositional possibilities that can fit into the apatite-like structure, which provide high compositional variations as corresponding to its non-stoichiometric character. Bottom; three different schemes and projections of the hydroxyapatite unit cell.

Biological apatites are calcium deficient; hence their Ca/P ratio is always lower than 1.67, which corresponds to a stoichiometric apatite. No biological hydroxyapatite shows a stoichiometric Ca/P ratio, but they all move towards this value as the organism ages, which are linked to an increase in crystallinity. These trends have a remarkable physiological meaning, since the younger, less-crystalline tissue can develop and grow faster, while storing other elements that the body needs during its growth; this is due to the highly nonstoichiometric

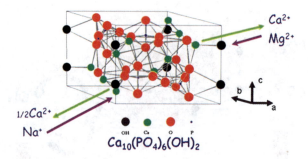

Figure 1.8   Likely substitutions in the cationic sublattice for biological apatites.

quality of HA, which caters for the substitutional inclusion of different amounts of several ions, such as Na^+, K^+, Mg^{2+}, Sr^{2+}, Cl^-, F^-, HPO_4^{2-}, *etc.*[21] (Figure 1.8).

Two frequent substitutions are the inclusion of sodium and magnesium ions in calcium lattice positions. When a magnesium ion replaces a calcium ion, the charge and position balance is unaffected. If a sodium ion replaces a calcium ion, however, this balance is lost and the electrical neutrality of the lattice can only be restored through the creation of vacancies, therefore increasing the internal disorder.

The more crystalline the HA becomes, the more difficult interchanges and growth are. In this sense, it is worth stressing that the bone is probably a very important detoxicating system for heavy metals due to the ease of substitution in apatites; heavy metals, in the form of insoluble phosphates, can be retained in the hard tissues without significant alterations of their structural properties.

However, the ability to exchange ions in this structure is not a coincidence. Nature designed it, and the materials scientist can use it as a blueprint to design and characterise new and better calcium phosphates for certain specific applications. It is known that the bone regeneration rate depends on several factors such as porosity, composition, solubility and presence of certain elements that, released during the resorption of the ceramic component, facilitate the bone regeneration carried out by the osteoblasts. Thus, for instance, small amounts of strontium, zinc or silicates stimulate the action of these osteoblasts and, in consequence, the new bone formation. Carbonate and strontium favour the dissolution, and therefore the resorption of the implant.[12] Silicates increase the mechanical strength, a very important factor in particular for porous ceramics, and also accelerate the bioactivity of apatite.[22] The current trend is, therefore, to obtain calcium phosphate bioceramics partially substituted by these elements. In fact, bone and enamel are some of the most complex biomineralised structures. The attempts to synthesise bone in the laboratory are devoted at obtaining biocompatible prosthetic implants, with the ability to leverage natural bone regeneration when inserted in the human body. Its formation might imply certain temporary structural changes on its components, which demand in turn the presence, at trace levels, of additional ions and molecules in order to enable the mineralisation process. This is the case, for instance, with bone growth processes, where the localised concentration of silicon-rich materials coincides precisely with areas of active bone growth. The reason is yet unknown, although the evidence is clear; the possible explanation of this phenomenon would also justify the great activity observed in certain silicon-substituted apatite phases and in some glasses obtained by sol-gel method, regarding cell proliferation and new bone growth.

1.1.2 A Discussion on Biomineralisation

Biomineralisation is the controlled formation of inorganic minerals in a living body; said minerals might be crystalline or amorphous, and their shape, symmetry and ultrastructure can reach high levels of complexity. Bioinorganic solids have been *replicated* with high precision throughout the evolution process, *i.e.* they have been reproduced identically to the primitive original. As a consequence, they have been systematically studied in the fields of biology and palaeontology. However, the chemical and biochemical processes of biomineralisation were not studied until quite recently. Such studies are currently providing new concepts in materials science and engineering.[17]

Biomineralisation studies the mineral formation processes in living entities. It encompasses the whole animal kingdom, from single-cell species to humans. Biogenic minerals are produced in large scale at the biosphere, their impact in the chemistry of oceans is remarkable and they are an important component in sea sediments and in many sedimentary rocks.

It is important to distinguish between mineralisation processes under strict biological – *genetic* – control, and those induced by a given biological activity that triggers a fortuitous precipitation. In the first case, these are crystal-chemical processes aimed at fulfilling specific biological functions, such as *structural support* (bones and shells), *mechanical rigidity* (teeth), *iron storage* (ferritin) and *magnetic* and *gravitational navigation,* while in the second case there are minerals produced with heterogeneous shapes and dimensions, which may play different roles in the *increase of cell density* or as means of *protection* against predators.[23]

At the nanometre scale, *biomineralisation* implies the molecular building of specific and self-assembled supramolecular organic systems (micelles, vesicles, *etc.*) which act as an environment, previously arranged, to control the formation of inorganic materials finely divided, of approximately 1 to 100 nm in size (Figure 1.9). The production of consolidated biominerals, such as *bones* and *teeth*, also requires the presence of previously arranged organic structures, at a higher length scale (micrometre).

The production of discrete or expanded architectures in biomineralisation frequently includes a hierarchical process: the building of organic assemblies made of molecules confers structure to the synthesis of arranged biominerals, which act in turn as preassembled units in the generation of higher-order complex microstructures. Although different in complexity, bone formation in vertebrates (support function) and shell formation in molluscs (protection

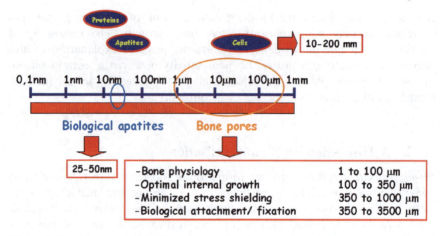

Figure 1.9 Scheme of the different scales for the most important hard-tissue-related biological moieties.

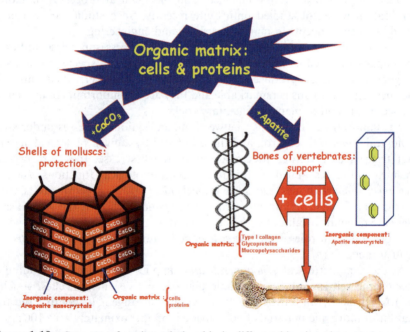

Figure 1.10 Structure–function relationship in different biominerals.

function) bear in common the crystallisation of inorganic phases within an organic matrix, which can be considered as a bonding agent arranging the crystals in certain positions in the case of bones, and as a bonding and grouping agent in shells (Figure 1.10).

Our knowledge of the most primitive forms of life is largely based upon the biominerals, more precisely in fossils, which accumulated in large amounts.

Several mountain chains, islands and coral reefs are formed by biogenic materials, such as limestone. This vast bioinorganic production during hundreds of millions of years has critically determined the development conditions of life.[24] CO_2, for instance, is combined in carbonate form, decreasing initially the greenhouse effect of the earth's crust. Leaving aside the shells, teeth and bones, there are many other systems that can be classified as biominerals: aragonite pellets generated by molluscs, the outer shells and spears of diatomea, radiolarian and certain plants, crystals with calcium, barium and iron content in gravity and magnetic field sensors formed by certain species, and the stones formed in the kidney and urinary system, although the latter are pathological biominerals. The protein ferritin, responsible for iron storage, can also be considered a biomineral, taking into account its structure and inorganic content.

Bones, horns and teeth perform very different biological functions and their external shapes are highly dissimilar. But all of them are formed by many calcium phosphate crystals, small and isolated, with nonstoichiometric *carbonate-hydroxyl-apatite* composition and structure, grouped together by an organic component. Nucleation and growth of the mineral crystals is regulated by the organic component, the *matrix*, segregated in turn by the cells located near the growing crystals (Figure 1.11).

This matrix defines the space where the mineralisation shall take place. The main components of the organic matrix are *cellulose*, in plants, *pectin* in

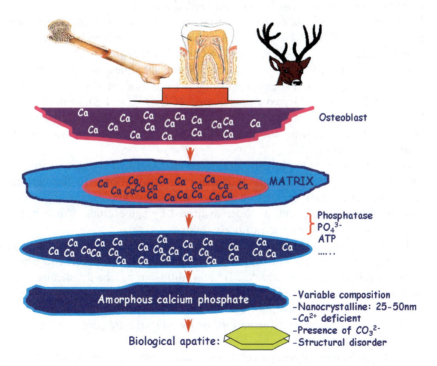

Figure 1.11 Calcium phosphate maturation stages during the formation of different mineralised structures.

diatomea, *chitin* and *proteins* in molluscs and arthropods, and *collagen* and *proteoglycans* in vertebrates.

1.1.3   Biomineralisation Processes

Different levels of biomineralisation can be distinguished, according to the type and complexity of the control mechanisms. The most primitive form corresponds to biologically *induced* biomineralisation, which is mainly present in bacteria and algae.[23] In these cases, biominerals are formed by spontaneous crystallisation, due to supersaturation provoked by ion pumps, and then polycrystalline aggregates are formed in the extracellular space. Gases generated in the biological processes, by bacteria for instance, can (and often do) react with metal ions from the environment to form biomineral deposits.

More complex mechanisms involve processes with higher biological control. The obtained, well-defined bioinorganic products are formed by *inorganic* and *organic* components. The organic phase is usually made of fibrous proteins, lipids or polysaccharides, and its properties will affect the resulting morphology and the structural integrity of the composite.

Whatever the case, the formation of an inorganic solid from an aqueous solution is achieved with the combination of three main physicochemical stages: *supersaturation*, *nucleation* and *crystal growth*.

Nucleation and *crystal growth* are processes that take place in a supersaturated medium and must be properly controlled in any mineralisation process. A living body is able to mineralise provided that there are well-regulated and active transport mechanisms available. Some examples of transport mechanisms are ion flows through membranes, formation or dissociation of ion complexes, enzyme-catalysed gas exchanges (CO_2, O_2 or H_2S), local changes of redox potential or pH, and variations in the medium's ionic strength. All these factors allow for creating and maintaining a supersaturated solution in a biological environment.[23]

Nucleation is related to kinetics of surface reactions such as *cluster* formation, growth of anisotropic crystals and phase transformations. In the biological world, however, there are certain surface structures that specifically avoid an unwanted nucleation, such as those exhibited by some kinds of fish in polar waters to avoid ice formation in body fluids.

The *growth* of a crystal or amorphous solid from a phase nucleus can be directly produced by the surrounding solution or by a continuous contribution of the required ions or molecules. Besides, diffusion can be drastically altered by any significant change in viscosity of said medium.

The controlled *growth* of biominerals can be also produced by a sequence of stages, through phase transformations or by intermediate precursors that lead to the solid-state phase.

Biomineralisation processes can be classified in two large groups; the first one includes those phenomena where it seems some kind of control exists over the mineralisation process, while the second one encompasses those where said control seems to be nonexistent. According to Mann,[23] biomineralisation processes can be described as *biologically induced* when said biomineralisation is due to the withdrawal of ion or residual matter from cells, and is verified in an open environment, *i.e.* not in a region purposely restricted. It is produced as a consequence of a slight chemical or physical disturbance in the system. The crystals formed usually give rise to aggregates of different sizes, with similar morphology to mineral inorganic crystals. Besides, the kind of mineral obtained depends on both the environmental conditions of the living organism and on the biological processes involved in the formation, since the same organism is able to produce different minerals in different environments. This is particularly so in single-cell species, although some higher-order species also verify this behaviour.

There are, however, situations where a specific mechanism is acting, which are then described as *biologically controlled*. An essential element of this process is the space localisation, whether at a membrane-closed compartment, or confined by cell walls, or by a previously formed organic matrix. The biologically controlled process of formation of biomaterials can be considered as the opposite to a biologically induced process. It is much more complex and implies a strict chemical and structural control.

Most biominerals formed under *controlled* conditions precipitate from solutions that are in turn *controlled* in terms of composition by the cells in charge; hence the contents of trace elements and stable isotopes in many mineralised areas are not balanced with the concentrations present in the initial medium.

Nucleation in controlled biomineralisation requires low supersaturation combined with active interfaces. Supersaturation is regulated by ion transport and processes involving reaction inhibitors and/or accelerators. The active interfaces are generated by organic substrates in the mineralisation area. Molecules present in the solution can directly inhibit the formation of nuclei from a specific mineral phase, hence allowing the growth of another phase.

Crystal *growth* depends on the supply of material to the newly formed interface. Low supersaturation conditions will favour the decrease in number of nuclei and will also restrict secondary nucleation, limiting somehow the disorder in the crystal phase. Under these low supersaturation conditions, growth rate is determined by the rate of ion bonding at the surface. In this scenario, foreign ions and large or small biomolecules can be incorporated to the surface, modifying the crystal growth and altering its morphology.

The final stage in the formation of a biomineral is its *growth interruption*. This effect may be triggered by a lack of ion supply at the mineralisation site, or because the crystal comes into contact with another crystal, or else because the mineral comes in contact with the previously formed organic phase.

Whatever the cause, biomineralisation processes are extremely complex, and not yet well known. One of the prevailing issues not yet fully elucidated is the mechanism at molecular level that controls the crystal formation process. If we consider the features of many organism-grown minerals, it seems that such control can be exerted at various levels. The lowest level of control would be exemplified by the less specific mineralisation phenomena, such as in many bacteria. These processes are considered more of an *induction* than a *control* of crystallisation. The opposite case would be the most sophisticated composites of crystals and organic matter, where apparently there is a total control on crystal *orientation* during its *nucleation*, and on its *size* and *shape* during the *growth stage*. This would be the case in the bones of vertebrates. In between these two examples we can find plenty of intermediate situations where some crystal parameters are controlled, but not all of them.[23]

A basic strategy performed by many organisms to control mineralisation is to seal a given space in order to regulate the composition of the culture medium. This is usually done forming barriers made of lipid bilayers or macromolecular groups. Subsequently, the sealed space can be divided in smaller spaces where individual crystals will be grown, adopting the shape of said compartment. An additional strategy is also to introduce specific acidic glycoproteins in the sealed solution, which interact with the growing crystals and regulate their growth patterns. There are many other routes to exert control, such as introducing ions at very precise intervals, eliminating certain trace elements, introducing specific enzymes, *etc.* All these phenomena are due to the activity of specialised cells that regulate each process throughout its whole duration.

The stereochemical and structural relationship between macromolecules from the organic matrix and from the crystalline phase is a very important aspect in the complex phenomenon of biomineralisation. These macro-molecules are able to control the crystal formation processes. It is already known that there is a wide range of biomineralisation processes in Nature, and that it is not possible to know *a priori* the specific mechanism of each one. It seems, however, that there are certain basic common rules regarding the control of crystal formation and the *interactions* involved. The term *interaction* refers here to the structure and stereochemistry of the phases involved, *i.e.* *nano-crystals* and *macromolecules*.

As already mentioned, the inorganic and organic components are forced to interact in order to produce a biomineral. They are not two independent elements; the specific extent and method for this interaction can be extremely varied, and the same variability applies to the biomineral's functionality.

1.1.4 Biominerals

The biominerals, natural composite materials, are the result of millions of years of evolution. The mineral phases present in living species can be also obtained in the laboratory or by geochemical routes. The synthesis conditions, however, are very different because the enforcement of said conditions at the biological environment is not so strict. It is worth noting that biogenic minerals usually differ from their inorganic counterparts in two very specific parameters: *morphology* and *order* within the biological system. It is quite likely that some general mechanisms exist that govern the formation of these minerals, and if our knowledge of these potentially general principles would improve, new options in material synthesis or modifications of already existing materials could be possible as an answer to a wide range of applications in materials science.

The biominerals or organic/inorganic composites used in biology exhibit some unique properties that are not just interesting *per se*; the study of the formation processes of these minerals can lead us to reconsider the world of industrial *composites*, to review their synthesis methods and to try and improve their properties. For instance, comparative studies with biominerals have provided new thinking on improvements of the physical-chemical properties in cements. In fact, the most noticeable property of minerals in biology is to provide *physical rigidity* to their host. But biological minerals are not just building material, as we could consider their role in shells, bones and teeth; they also fulfil many other purposes such as, for instance, in sensing devices. The biomineralisation process is responsible for bone formation, growth of teeth, shells, eggshells, pearls, coral and many other materials that form part of living species. *Biomineralisation* is hence responsible for the controlled formation of minerals in living organisms. These biominerals can be either crystalline or amorphous, and they belong in the *bioinorganic* family of solids. Bioinorganic solids are usually a) remarkably nonstoichiometric, that is, with frequent variations in their composition, allowing impurities to be included as *interstitial* and/or *substitutional* defects, b) they can be present in amorphous and/or crystalline form, and in some situations several polymorphs of the same crystalline solid can coexist. Besides, the inorganic component is just a part of the resulting biomineral that actually is a composite material, or more precisely a *nanocomposite,* formed by an organic matrix which restricts the growth of the inorganic component at perfectly defined and delimited areas in space, determining a strict shape and size control.

The organic component might be a vesicle, perhaps a protein matrix; whatever the case, biominerals are formed by very different chemical systems, since they require the combined participation of *mineral components* and *organic molecules*. Vesicles give rise to *three-dimensional* structures, and are able to fill cavities, while the organic molecules can form *linear* or *layered* structures, and also can interact with the inorganic matrix, generating the voids to be filled with minerals.

Almost half of the biominerals known include the element *calcium* among their constituents. This is the reason why the term *calcification* is often used to describe the processes where an inorganic material is produced by a living organism. But this generalisation is not always true, since there are many biominerals without any calcium content. Therefore, the term biomineralisation is not only much more generic but also more adequate, encompassing all inorganic phases regardless of their composition; the outcome is the biomineral, that is, a mineral inside a living organism, which is a truly *composite material*.

Biomineralisation processes give rise to many inorganic phases; the four most abundant are *calcite, aragonite, apatite* and *opal*.

In load-bearing biominerals, such as bones, some stress-induced changes may appear and induce in turn certain consequences on their properties, in the crystal growth for instance. The *growth* of biominerals is related to one of the great unsolved issues in biology: *the morphology of its nano- or microcrystals*. The skeletons of many species exhibit peculiarities that are clearly a product of their morphogenesis, with direct effects on it, since the gametes of biological systems never or hardly ever produce a biomineral precipitate.

Another question to be considered is the relevance of biominerals from a *chemical* point of view. Many of these minerals act as deposits that enable to regulate the presence of cations and free anions in cell systems. Concentrations of iron, calcium and phosphates, in particular, are strictly controlled. A biomineral is the best possible *regulator* of homeostasis. It is important to recall that exocytosis of mineral deposits is a very simple function for cells, enabling to eliminate the excess of certain elements. In fact, some authors believe that calcium metabolism is mainly due to the need to reject or eliminate calcium excess, leading to the development and temporary storage of this element in different biominerals. However, some evidence counters the validity of this point of view: many living species build their skeletons with elements that do not have to be eliminated, such as silicon.[25]

Mineral deposits such as iron and manganese oxides are used as *energy sources* by organisms moving from oxic to anoxic areas. Therefore, biominerals are also used by some living species as an *energy source* to carry out certain biological processes. This fact has been verified in marine bacteria.[26]

Although silicon – in silicate form – is the second most abundant element in the Earth's crust, it plays a minor part in the biosphere. It may be due in part to the low solubility of silicic acid, H_4SiO_4, and of amorphous silica, $SiO_n(OH)_{4-2n}$. In an aqueous medium, at pH between 1 and 9, its solubility is approximately 100–140 ppm. In presence of cations such as calcium, aluminium or iron, the solubility markedly decreases, and solubility in sea water is just 5 ppm. At the biosphere, amorphous silicon is dissolved and then easily reabsorbed in the organism; it will then polymerise or connect with other solid structures.

Amorphous silicon biomineral is mainly present in single-cell organisms, in silicon sponges and in many plants, where it is located in fitolith form at cell membranes of grain plants or types of grass, with a clear deterrent purpose. The fragile tips of stings in some plants, such as nettles, are also made of amorphous silicon.

There is a wide range of biological systems with biomineral content, from the human being to single-cell species. Modern molecular biology indicates that single-cell systems may be the best object of research in order to improve our knowledge of a biological structure.

1.1.5   Inorganic Components: Composition and Most Frequent Structures

At present, there is a wide range of known inorganic solids included among the so-called biominerals. The main metal ions deposited in single-cell or multiple-cell species are the divalent alkali-earth cations Mg, Ca, Sr, Ba, the transition metal Fe and the semimetal Si. They usually form solid phases with anions such as carbonate, oxalate, sulfate, phosphate and oxides/hydroxides. The metals Mn, Au, Ag, Pt, Cu, Zn, Cd and Pb are less frequent and generally deposited in bacteria, in sulfide form. More than 60% of known minerals contain hydroxyl groups and/or water bonds, and are easily dissolved releasing ions. The crystal lattice of the mineral group including metal phosphates is particularly prone to inclusions of several additional ions, such as fluorides, carbonates, hydroxyls and magnesium. In some cases, this ability allows for the modification of the material's crystal structure and hence of its properties.

The field of biominerals encompasses a wide range of inorganic salts with many different functions, which are present in several species in Nature. For instance, calcium in carbonate or phosphate form is important for nearly all the species, while calcium sulfate compounds are essential for very few species. All along the evolution of species, there has been a constant development of the *control of selective precipitation*, that is, of *nucleation* and *growth processes*, as well as the *shape of the precipitates* and their exact *location within a living body*.[27]

The minerals in structures aimed at providing support or external protection can be crystalline or amorphous. The generation of amorphous materials in any kind of biological system is undoubtedly a favourable process from an energetic perspective, and is present in several examples such as carbonates and biological phosphates. This *amorphous phase* usually leads to a series of transformations, either as consequence of *recrystallisation* processes – which give rise to a crystalline phase, likely to transform itself into other phases due to *in-situ* structural modifications – or due to redissolution of the amorphous phase, enabling the *nucleation* of a new phase. If the minerals are crystalline, the biological control can be exerted over several parameters: *chemical composition*, *polymorph formation*, and *crystal size and shape*. Each one of these parameters is in turn closely related to the organic matrix controlling *elements concentration*, crystal *nucleation* and *growth*. If the mineral is amorphous, the chemical composition allows for almost infinite variations, although a certain concentration of the essential elements remains crucial. A typical amorphous biomineral is hydrated silica, $SiO_n(OH)_{4-2n}$, where n can be any value in the range from 0 to 2. Several forms of hydrated silica can be found in living organisms, both in the *sea world* – such as sponges, diatomea, protozoa and single-cell algae – and in the *vegetable kingdom*, present in *amorphous* form. The actions performed by these species to mineralise silicic acids are extremely complex. It seems that this process first involves the transportation of silicic acid towards the inside of the cell, and then to the deposition locations where the monomer will be polymerised to silica. For any silicon structure to be generated, the preliminary essential requirement is the availability of silicic acid, which must also be transported in adequate concentrations. If this stage is verified, the nucleation and polymerisation processes may begin, which will eventually lead to the development of strict and specific morphological features, both at the microscopic and macroscopic scales. Little is known about the early stages, previous to deposition. There are several mechanisms that have been suggested to try to explain biosilication, but none of them is conclusive yet.

Some biominerals perform a very specific function within the biological world; they work as sensors, both for positioning and attitude or orientation. The inorganic minerals generated by some species to carry out this task are *calcite, aragonite, barite* and *magnetite*.

1.1.6 Organic Components: Vesicles and Polymer Matrices

The most common cell organ is the *vesicle*.[27] It is an aqueous compartment surrounded by a lipid membrane, impervious to all ions and most organic molecules. The ions required to form the biomineral are accumulated in the vesicle by a pumping action. These ions are, among others, Ca^{2+}, H^+, SO_4^{2-}, HPO_4^{2-}, HCO_3^-. In order to understand the biomineral formation, a great deal will depend on the knowledge of cell vesicles and ion pumps.

Proteins or *polysaccharides* are able to build another kind of *receptacle, mould* or *sealed container*, more or less impervious to ions and molecules, depending on the particular system. This receptacle might be into the cell itself, as in the case of ferritin, or outside the cell, such as bone collagen for instance. The exact shape of the protein receptacle for ferritin is fixed, and also the open spaces in collagen where apatite grows always exhibit the same shape. In contrast, the available space in a typical vesicle is not controlled by the organic structure, since vesicles do not have internal crosslinks in their membranes. In fact, vesicle space is very different from cytoplasmatic space, which usually includes crossed-fibre structures. As a consequence, when the *mould* is made of protein or polysaccharides, precipitation must be controlled through the regulation of cytoplasmatic or extracellular homeostasis. Extracellular fluids have a sustained chemical composition due to the actions of control organs such as the kidney, which actually works as a macropump.

Most of the controlled mineralisation processes performed by organisms exhibit associated macromolecules. These macromolecules carry out important tasks in tissue formation and modification of the biomechanical properties of the final product. Although there are thousands of different associated macromolecules, Williams[27] stated that they all can be classified in two types: structural *macromolecules* and *acid macromolecules*. The main structural macromolecules are *collagen, α-* and *β-quitine*, and *quitine-protein complexes*. The main acid macromolecules are not very well defined in some organisms, but we may include in this group *glycoproteins, proteoglicans, Gla-rich proteins*, and *acid polysaccharides*. Little is known about the secondary conformation of acid macromolecules, apart from the fact that all acid glycoproteins with high contents of glutamic and aspartic acids partially adopt *in vitro* the β layer conformation, in the presence of calcium. Although the composition of these macromolecules shows little variations between species, the opposite can be said of structural macromolecules. They vary from one tissue to another, and there are even some hard mineralised pieces that do not *seem* to have any kind of acid macromolecule at all. This lack of presence in some tissues allows us to infer that their purpose might be to modify the mechanical properties of the final product, not to regulate biomineralisation. The main means of control over biominerals are the independent areas in the cytoplasmatic space or in the extracellular zones in multiple-cell species, where the organic structures develop well-defined volumes and external shapes.

There are different physical and chemical controls in the development of a mineral phase. Physical controls are determined by the physics of our world and by biological source fields, in the same way that biological chemistry is restricted by the properties of chemical elements in the periodic table.

Mineral and vesicle grow together under the influence of many macroscopic fields. It should be taken into account that the functional values often depend on the interactions with these fields, due to the density, magnetic properties, ion mobility in the crystal lattice, elastic constants and other material properties. These properties do not fall under a strict biological control. Microscopic shape is restricted by the rules of symmetry in crystalline materials, but not in amorphous ones. Any crystal-based biomineral exhibits many restrictions in shape, and the organism adapts itself to them.[23]

1.2 Alternatives to Obtain Nanosized Calcium-Deficient Carbonate-Hydroxy-Apatites

Hydroxyapatite, (HA), $Ca_{10}(PO_4)_6(OH)_2$ is the most widely used synthetic calcium phosphate for the implant fabrication because is the most similar material, from the structural and chemical point of view, to the mineral component of bones.[28] HA with hexagonal symmetry S.G. $P6_3/m$ and lattice parameters $a = 0.95$ nm and $c = 0.68$ nm, exhibits excellent properties as a biomaterial, such as biocompatibility, bioactivity and osteoconductivity. When apatites aimed to mimic biological ones are synthesised, the main characteristics required are small particle size, calcium deficiency and the presence of $[CO_3]^{2-}$ ions in the crystalline network. Two different strategies can be applied with this purpose.

The first one is based in the use of chemical synthesis methods to obtain solids with small particle size. There are plenty of options among these wet-route processes, which will be generally termed as the *synthetic route*.[29]

The other strategy implies the collaboration of physiological body fluids.[30] In fact, certain ceramic materials react chemically with the surrounding medium when inserted in the organism of a vertebrate, yielding biological-like apatites through a process known as the *biomimetic process*.

1.2.1 The Synthetic Route

Some synthetic strategies used to obtain submicrometric particles are the aerosol synthesis technique,[31] methods based on precipitation of aqueous solutions,[32,33] or applications of the sol-gel method, or some of its modifications such as the liquid mix technique, which is based on the Pechini patent.[34,35] In these methods, the variation of synthesis parameters yields materials with different properties. Quantum/classical molecular mechanics simulations have been used to understand the mechanisms of calcium and phosphate association in aqueous solution.[36]

On the other hand, it is difficult to synthesise in the laboratory calcium apatites with carbonate contents analogous to those in bone. Indeed, it is difficult to avoid completely the presence of some carbonate ions in the apatite network, but the amount of these ions is always inferior to that in natural bone values (4–8 wt%) and/or they are located in different lattice positions.[12,37] It must be taken into account that biological apatites are always of type B, but if the synthesis of the ceramic material takes place at high temperatures, type-A apatites are obtained. Synthesis at low temperatures allows apatites to be obtained with carbonate ions in phosphate positions but in lower amounts than in the mineral component of bones.[38,39]

1.2.2 The Biomimetic Process

As in any other chemical reaction, the product obtained when a substance reacts with its environment might be an unexpected or unfavourable result, such as corrosion of an exposed metal, for instance, but it could also lead to a positive reaction product that chemically transforms the starting substance into the desired final outcome. This is the case of *bioactive ceramics*, which chemically react with body fluids towards the production of newly formed bone. When dealing with the repair of a section of the skeleton, there are two different basic options to consider: *replacing* the damaged part, or *substituting* it and regenerating the bone tissue. This is the role played by bioactive ceramics.[40]

Calcium phosphates, glasses and glass ceramics, the three families of ceramic materials where several bioactive products have been obtained, have given rise to starting materials used to obtain mixtures of two or more components, in order to improve its bioactive response in a shorter period of time.

These types of ceramics are also studied to define shaping methods allowing implant pieces to be obtained in the required shapes and sizes, with a given porosity, according to the specific role of each ceramic implant. Hence, if the main requirement is to verify in the shortest possible time a chemical reaction leading to the formation of nanoapatites as precursors of newly formed bone, it will be necessary to design highly porous pieces, which must also include a certain degree of macropores to ensure bone oxygenation and angiogenesis.

However, these requirements are often discarded when designing the ceramic piece. As a result, the chemical reaction only takes place on the external surface of the piece (if made of bioactive ceramics) or it simply does not occur if the piece is made of an inert material; in both cases, the inside of the piece remains as a solid monolith able to fulfil bone replacement functions, but without the regenerative role associated to bioactive ceramics. In order to achieve a chemical reaction throughout the whole material, it is important to design pieces with bone-like hierarchical structure of pores. In this way, the fluids will be in contact with a much larger specific surface, reaching a higher reactivity phase that allows full reaction between the bioactive ceramic and the fluids to be achieved, thus yielding newly formed bone as reaction product.

References

1. M. Vallet-Regí and J. González-Calbet, *Prog. Solid State Chem.*, 2004, **32**, 1.
2. S. Mann, J. Webb and R. J. P. Williams, *Biomineralization. Chemical and Biochemical Perspectives*, VCH, Weinheim, Germany, 1989.
3. D. Lee and M. J. Glimcher, *J. Mol. Biol.*, 1991, **217**, 487.
4. A. J. Friedenstein, *Int. Rev. Cytol.*, 1976, **47**, 327.
5. M. J. Glimcher, In *Disorders of Bone and Mineral Metabolism*, F. L. Coe and M. J. Favus, eds., Raven Press, New York, 1992, 265–286.
6. M. J. Glimcher, In *The Chemistry and Biology of Mineralized Connective Tissues*, A. Veis, ed., Elsevier, Amsterdam, 1981, 618–673.
7. L. T. Kuhn, D. J. Fink and A. H. Heuer, *Biomimetic Strategies and Materials Processing*, In *Biomimetic Materials Chemistry*, Stephen Mann, ed., Wiley-VCH, United Kingdom, 1996, 41–68.
8. S. P. Bruder, A. I. Caplan, Y. Gotoh, L. C. Gerstenfeld and M. J. Glimcher, *Calcif. Tissue Int.*, 1991, **48**, 429.
9. M. D. McKee, A. Nanci, W. J. Landis, Y. Gotoh, L. C. Gertenfeld and M. J. Glimcher, *Anat. Rec.*, 1990, **228**, 77.
10. Y. Gotoh, L. C. Gerstenfeld and M. J. Glimcher, *Eur. J. Biochem.*, 1990, **228**, 77.
11. J. C. Elliott, *Structure and Chemistry of the Apatites and other Calcium Orthophosphates*, ed., Elsevier, London, 1994.
12. R. Z. LeGeros, In: *Monographs in Oral Science, Vol. 15: Calcium Phosphates in Oral Biology and Medicine*, H. M. Myers and S. Karger, ed. Basel, 1991.
13. D. G. Pechak, M. J. Kujawa and A. I. Caplan, *Bone.*, 1986, **7**, 441.
14. E. D. Eanes and J. L. Meyer, *Calcif. Tissue Res.*, 1977, **23**, 259.
15. H. Nancollas, *In vitro Studies of Calcium Phosphate Crystallization. In Biomineralization. Chemical and Biochemical Perspectives*, S. Mann, J. Weobb, R. J. P. Williams, ed., VCH, Weinheim, Germany, 1989, 157–188.
16. A. Veis, *Biochemical Studies of Vertebrate Tooth Mineralization*, In *Biomineralization. Chemical and Biochemical Perspectives.*, S. Mann, J. Webb and R. J. P. Williams, eds., VCH, Weinheim, Germany, 1989,189–222.
17. J. D. Birchall, *The Importance of the Study of Biominerals to Materials Technology*, In *Biomineralization. Chemical and Biochemical Perspectives.*, S. Mann, J. Webb and R. J. P. Willians, eds., VCH, Weinheim, Germany, 1989, 491–508.
18. J. B. Park and R. S. Lakes, *Structure-Property Relationships of Biological Materials*, In *Biomaterials. An Introduction*, ed., Plenum Press, New York and London, 1992, 185–222.

19. J. B. Park and R. S. Lakes, *Composites as Biomaterials*, In *Biomaterials. An Introduction*, 2nd Edn., Plenum Press, New York and London, 1992, 169–183.
20. J. Christofferson and W. J. Landis, *Anat. Rec.*, 1991, **230**, 435.
21. M. Vallet-Regí, *Anales de Quím. Inter.*, 1 ed., Suplement 1. 1997, **93**.1, S6.
22. M. Vallet-Regí and D. Arcos, *J. Mater. Chem.*, 2005, **15**, 1509.
23. S. Mann, *Crystallochemical Strategies in Biomineralization*, In *Biomineralization. Chemical and Biochemical Perspectives*, S. Mann, J. Webb and R. J. P. Williams, eds., VCH, Weinheim, Germany, 1989, 35–62.
24. M. A. Borowitzka, *Carbonate Calcification in Algae-Initiation an Control*, In *Biomineralization. Chemical and Biochemical Perspectives*, S. Mann, J. Webb and R. J. P. Williams, eds., VCH, Weinheim, Germany, 1989, 63–94.
25. C. C. Perry, *Chemical Studies of Biogenic Silica*, In *Biomineralization. Chemical and Biochemical Perspectives*, S. Mann, J. Webb and R. J. P. Williams, eds., VCH, Weinheim, Germany, 1989, 223–256.
26. S. Mann and R. B. Frankel, *Magnetite Biomineralization in Unicellular Microorganisms*, In *Biomineralization. Chemical and Biochemical Perspectives. S. Mann*, J. Webb and R. J. P. Williams, eds., VCH, Weinheim, Germany. 1989. 389–426.
27. R. J. P. Williams, *The Functional Forms of Biominerals*, In *Biomineralization. Chemical and Biochemical Perspectives.*, S. Mann, J. Webb and R. J. P. Williams, eds., VCH, Weinheim, Germany, 1989, 1–34.
28. M. Vallet-Regí, *J. Chem. Soc. Dalton Trans.*, 2001, 97.
29. M. Vallet-Regí, *Preparative Strategies for controlling structure and morphology of metal oxides*, In *Perspectives in Solid State Chemistry.*, K. J. Rao eds., Narosa Publishing House, India, 1995, 37–65.
30. M. Vallet-Regí, C. V. Ragel and A. J. Salinas, *Eur. J. Inor. Chem.*, 2003, **6**, 1029.
31. M. Vallet-Regí, M. T. Gutiérrez-Ríos, M. P. Alonso, M. I. de Frutos and S. Nicolopoulos, *J. Solid State Chem.*, 1994, **112**, 58.
32. M. Vallet-Regí, L. M. Rodríguez Lorenzo and A. J. Salinas, *Solid State Ionics.*, 1997, **101–103**, 1279.
33. L. M. Rodríguez-Lorenzo and M. Vallet-Regí, *Chem. Mater.*, 2000, **12(8)**, 2460.
34. M. P. Pechini, (July 11, 1967) U. S. Patent 3,330,697; 1967.
35. J. Peña and M. Vallet-Regí, *J. Eur. Ceram. Soc.*, 2003, **23**, 1687.
36. D. Zahn, *Z. Anorg. Allg. Chem.*, 2004, **630**, 1507.
37. J. C. Elliot, G. Bond and J. C. Tombe, *J. Appl. Crystallogr.*, 1980, **13**, 618.
38. M. Okazaki, T. Matsumoto, M. Taira, J. Takakashi and R. Z. LeGeros, In: *Bioceramics, Vol. 11*, R. Z. LeGeros, eds., World Scientific, New York, 1998, 85.
39. Y. Doi, T. Shibutani, Y. Moriwaki, T. Kajimoto and Y. J. Iwayama, *J. Biomed Mater. Res.*, 1998, **39**, 603.
40. A. J. Salinas and M. Vallet-Regí, *Z. Anorg. Allg. Chem.*, 2007, **633**, 1762.

CHAPTER 2
Synthetic Nanoapatites

2.1 Introduction

2.1.1 General Remarks on the Reactivity of Solids

The most common reactions that a chemist needs to know in order to obtain a solid are those starting from two reactants in solution, leading to a new compound that is insoluble in the solvent used, usually water. There are, however, many other types of reactions that also lead to the synthesis of a solid (Figure 2.1). The main difference between classical synthesis from a solution and all the other synthesis routes depicted in the figure is the lack of a solvent, *i.e.* of an easy transport medium for the reactants, although its presence imposes a restriction on the feasible temperature range for the reaction, since it cannot exceed the boiling point of said solvent.

According to Figure 2.1, it is possible to obtain solids from reactants in solid, melted or even gaseous state, increasing remarkably the temperature range available; this fact allows us to prepare solids that would be otherwise unfeasible by a conventional method. These principles can be directly applied to the laboratory synthesis of apatites. Although there are obvious differences between the four alternative routes depicted above, which can be even more complex if the reactants themselves are in dissimilar phases (liquid/solid, gas/solid, *etc.*), the common feature in all these processes is the synthesis and outcome of a new phase. This means that a new interface has appeared, with associated thermodynamical restrictions to its formation (nucleation stage), which are not present in homogeneous processes. Besides, the wider temperature range associated with solvent-free synthesis, while being clearly an advantage, does impose remarkable restrictions from a kinetic point of view on solid–solid synthesis reactions. This process is determined by the low mobility of the reactants.

RSC Nanoscience & Nanotechnology
Biomimetic Nanoceramics in Clinical Use: From Materials to Applications
By María Vallet-Regí and Daniel Arcos

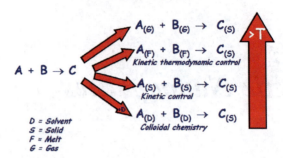

Figure 2.1   Scheme of possible reactions that lead to solid product formation.

Solid formation reactions are usually classified in five groups:

1. Solid → products
2. Solid + gas → products
3. Solid + solid → products
4. Solid + liquid → products
5. Surface reactions in solids.

The first group includes *decomposition* of solids and *polymerisation*. The second group corresponds to *oxidation* or *reduction* reactions. The solid–solid reactions of the third group take place for instance in the *ceramic* method, the most traditional synthesis method in the world of cements and ceramic materials. The fourth group includes reactions such as intercalation and percolation, while the fifth group holds all those reactions occurring in the surface of solids.

Solid-state reactions may include one or more elementary stages such as *adsorption* or *desorption* of gas phases onto the solid surface, chemical reactions at the *atomic scale*, *nucleation* of a new phase and *transportation* phenomena through the solid. Besides, external factors such as *temperature, surrounding environment, irradiation, etc.*, significantly affect the reactivity.

There are multiple factors influencing the reactivity of solids. In fact, features such as *particle size, gas atmosphere* and *external additives*, as well as *dopants* and *impurities*, play a predominant role in reactivity. Reactivity, for instance, increases when the particle size decreases. In this sense, also the use of solid reactants with small particle size leads to more homogeneous solid products. The atmosphere where the reaction takes place has clear effects on the reactivity, even more if the gas is also an exchangeable component of the solid phases. Doping with certain species also determines the reaction kinetics. And impurities lower the temperature required for a given reaction.

According to these observations, it seems clear that the *previous history* of any solid is extremely important for its future reactivity. The preparation method used may have determined a certain particle size, impurities, defects, which forcefully affect the subsequent reactivity of this solid. A mechanical treatment in a mortar or ball mill, for instance, greatly affects the treated solid, creating different types of defects that may determine the kinetic of the whole process.

The synthesis of *tailored solids,* with predetermined structure and properties, is the main and most difficult challenge in solid-state chemistry, which plays a crucial role in the fields of materials science and technology.

In the last few years, scientists working in solid-state chemistry have put special efforts into the study and development of new synthesis methods. Due to the vast number of theoretically possible solids to be obtained, the synthesis tools to be used may vary with the issues to be solved in each particular case. Luckily, at present, there are adequate techniques available to control both the *structure* and *morphology* of many different materials. A well-designed synthesis process does require in all cases a profound knowledge of *crystallochemistry* together with a good control of the particular *thermodynamics, phase diagram* and reaction *kinetics* involved; all this, added to the information available in literature, is the first and vital step in the design and synthesis of new apatites with tailored properties.

Theoretically speaking, it is possible to design the *properties* using the classical tools: control of *structure* and *composition*. Besides, the properties of apatites are closely related with their *previous history*; it is important to choose carefully the synthesis method and to carry out a detailed microstructural characterisation in order to correlate the influence of structure and defects on its properties.

2.1.2 Objectives and Preparation Strategies

In order to modify the properties of apatites, two strategies may be followed:

a) To produce *structural changes* preserving its chemical composition.
b) To introduce *compositional changes* avoiding changes in the average structure.

The latter may allow a systematic search of *new compositions* to obtain new and better properties, hence *designing* tailored apatites.

A valid motto for a solid-state chemist would be "to understand all available synthesis methods to obtain a given solid, in order to always choose the optimum one". This is the strategy that has to be applied with apatites, using different synthesis methods and opening up new expectations in the field of applications.

In the words of Prof. C.N.R. Rao,[1] *it is useful to distinguish between synthesis of new solids and synthesis of solids by new methods*. To obtain a new solid, it is not always compulsory to apply a new method. It could be very useful, however, to synthesise already known materials using different routes that allow modification of their *texture* and *microstructure*.

There are plenty of methods nowadays to obtain apatites. Once again, it is very important to establish first our objectives, before initiating a synthesis process.

2.2   Synthesis Methods

Synthesis of apatites from solid precursors implies slow solid-state reactions, and it is usually difficult to achieve a complete reaction. Long treatment periods and high temperatures are needed, in order to improve the diffusion of the atoms implied in the reaction throughout their respective solid precursors, reaching the interface where the reaction is actually happening. It is also possible to produce a solid-state transformation from a given phase to another one with equal composition, whether under high temperature or pressure, or under a combination of both.

The synthesis of solids from liquids occurs by solidification of the melted product, obtaining *single crystals* when the cooling rate is low enough, or noncrystalline materials, *glasses*, when the cooling rate is high enough as to avoid the ordered arrangement of atoms, and hence crystallisation. This is not the most common way to obtain apatites. There is a more adequate alternative for apatite synthesis, namely crystallisation of solids from solutions. It is rather frequent that a solid is obtained from a liquid phase, where the formation of the solid product is a purely physical process and corresponds to a phase transformation. In other cases, the synthesis incorporates a liquid. These synthesis routes may be classified according to the quality of *melted matter* or *solution* of the *precursor liquid phase*.

Synthesis of solids from condensation of reactants in gaseous phase gives rise usually to solids in the form of *thin films* deposited onto adequate substrates.

Obviously, several techniques have been utilised for the preparation of hydroxyapatite and other calcium phosphates,[2-4] which include precipitation, hydrothermal and hydrolysis of other calcium phosphates.[5-36] Modifications of these "classical" methods (precipitation, hydrolysis or precipitation in the presence of urea, glycine, formamide, hexamethylenetetramine . . . .)[37-41] or alternative techniques have been employed to prepare hydroxyapatite with morphology, stoichiometry, ion substitution or degree of crystallinity as required for a specific application. Among them, sol-gel,[42-51] microwave irradiation,[52,53] freeze-drying,[54] mechanochemical method,[55-59] emulsion processing,[60-62] spray pyrolysis,[63-65] hydrolysis of α-TCP,[66] ultrasonics,[67,68] *etc.*, can be outlined.

2.2.1   Synthesis of Apatites by the Ceramic Method

The most traditional method in apatite synthesis is the *ceramic* method, which consists in a solid–solid reaction where both reactants and products are in the solid state. The usual starting phases are oxides, carbonates or, generally speaking, salts, with very different particle sizes and irregular morphologies (Figure 2.2). When mixed and homogenised in the stoichiometric ratio, they are subsequently submitted to an adequate thermal treatment to start the reaction. In most cases, this method requires high temperatures and long heating periods.

The study of chemical reactions between solid materials is a fundamental aspect of solid-state chemistry, allowing the influence of structure and defects in the reactivity of solids to be understood. It is important to determine which

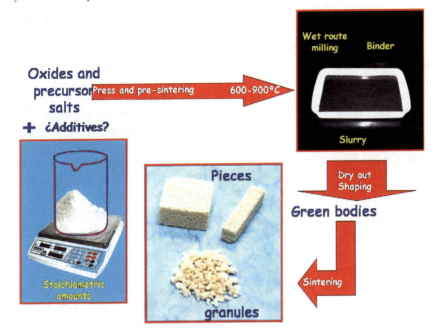

Figure 2.2 Scheme of the ceramic method.

factors rule the reactivity in the solid state, in order to obtain new solids with the desired structure and properties.

Solid-state reactions differ in a fundamental aspect from those taking place at a homogeneous fluid medium; the intrinsic reactivity of liquid or gaseous state reactions mainly depends on the intervening chemical species and their respective concentrations, while solid-state reactions greatly depend on the crystallinity of the chemical constituents. The fact that said constituents (atoms, ions or molecules) occupy fixed positions in determined sites of a given crystal lattice brings a new dimension to the reactivity of solids, in contrast with other physical states.

In other words, the chemical reactivity of solids is often more determined by the crystalline structure and the presence of defects, than by the intrinsic chemical reactivity of its constituents. This fact is clearly evidenced in a type of solid-state reaction termed *topochemical* or *topotactical*.

Another type of solid-state reaction occurs in *intercalation* processes. Also in *catalytic* reactions, or in many fields where catalysis plays a fundamental role, it is worth considering that not only the *crystal order* of the chemical constituents is important, but also their *particle size*. This is somehow implicit when considering that the chemical reactivity of solids relies mainly on their *crystalline structure* and *defects,* since the surface of any crystal particle can be considered as a plane defect. The smaller the particle, the lower the number of complete unit cells forming the crystal; as a consequence, the constituents will have a short *diffusion route*, reaching higher levels of reactivity. Another important factor is that the specific surface of small particles is significantly higher, and all these reasons combined confirm the importance of particle size in the reactivity of solids.

Solid-state reactions begin by interphase reactions at the contact points between the reactants. The product phase represents a kinetic obstacle for the ongoing reaction, which keeps reacting due to the diffusion of the constituents, which again come into contact. These difficulties often lead to the impossibility to obtain a pure single-phase, homogeneous product by this procedure (Figure 2.3).

The ceramic method is perhaps the best example to understand the reactions between solids. As previously mentioned, in liquid or vapour state the reacting molecules have more opportunities for contact between them and to react due to the continuous movement under conditions determined by statistical laws. To put it simply, diffusion is extremely easy in these two media.

In the solid state, on the contrary, the reactions generally take place between apparently regular crystalline structures, where the movements of the constituent species are much more restricted and depend to a complex degree on the presence of defects. Besides, said interaction can only occur at points of close contact between the reacting phases. Moreover, another difference is that in liquid-state reactions, the formed product does not affect greatly the course of the reaction.

However, in the solid state, the production of a more or less static layer of product can inhibit or at least slow down the progress of the reaction; it cannot carry on without contact, hence the diffusion albeit restricted, is the only way in which the reaction can continue in the solid state.

The study of reactions between solids could be expected to be simple, since they occur in a homogeneous state of the matter, as is the case in processes taking place with all elements in liquid or gas phase; the truth, however, is quite different. Reaction processes between solid precursors are extremely complex, dealing with two or more reacting phases plus the reaction product(s). There are difficult theoretical and experimental issues in this study.

When studying reactions in solids and chemical reactions in general, it is important to distinguish between *thermodynamical* and *kinetic* issues; among them, it is worth considering those factors that may improve the reaction kinetics.

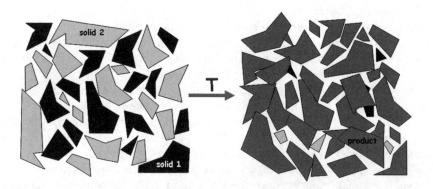

Figure 2.3 Scheme representing the solid reactants heterogeneity. Complete reaction is not attained due to the kinetic characteristics of the process.

Since the chemical potential and the activity of a pure solid remain constant at constant temperature and pressure, ΔG is an invariant for a given reaction process. When $\Delta G > 0$, the reaction does not start spontaneously. On the other hand, when $\Delta G < 0$, the reaction should be produced spontaneously. Even under these latter thermodynamically favourable conditions, solid-state reactions may not be completed due to the formation of a reaction product layer at the interphase area, which becomes larger as the reaction progresses; at least one of the reactants must cross this layer in order to continue the reaction.

According to the first two principles of thermodynamics $\Delta G = \Delta H - T\Delta S$, where ΔH and ΔS are the enthalpy and entropy variations during the reaction. In most cases, solid-state reactions imply a regrouping of the crystalline lattice as evidenced in many examples. The degree of crystallinity does not vary much in these cases; hence ΔS always has a value close to zero. A reaction will be verified if $\Delta H < 0$, because ΔG forcefully adopts a negative value ($\Delta G < 0$). Therefore, solid-state reactions are usually exothermic.

Generally speaking, the theoretical study of reaction *mechanisms* is carried out using a geometrical model that considers a *counterflow diffusion* of the various species forming both solids (Figure 2.4).

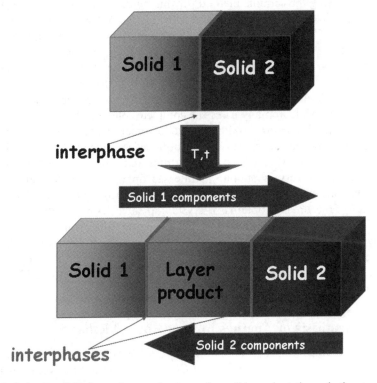

Figure 2.4 Possible formation mechanism of a solid product through the reactants interface.

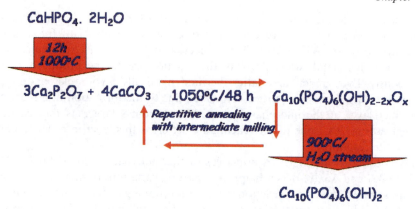

Figure 2.5 Ceramic method for hydroxyapatite synthesis.

   The production of hydroxyapatite by the ceramic method is a traditional laboratory process, which can be carried out using different precursor salts from the phosphate and carbonate families, respectively.

   Figure 2.5 depicts a possible route of solid-state reaction used in the laboratory synthesis of crystalline hydroxyapatite similar to mineral apatites available in Nature. Using the ceramic method for this synthesis, the starting precursors exhibit a particle size of around $10 \, \mu m$, which is the order of magnitude of the particle size of any chemical salt-type compound in commercial form. Ten micrometres may roughly correspond to a succession of 10 000 unit cells, which is the diffusion route to be covered by ions of each one of the reactants in order to reach the interphase of the other reactant, so that the chemical reaction can take place. Therefore, the kinetic obstacles greatly restrict the ionic diffusion, rendering impossible in many cases the complete solid–solid chemical reaction. All cases require starting products in stoichiometric amounts, submitted to milling and homogenising processes. The obtained starting mixture must then be submitted to generally very high temperatures and long treatment times. This procedure allows very crystalline apatites as opposed to biological apatites, where the particle size ranges between 25–50 nm, to be obtained (Figure 2.6). Therefore, if the aim is to obtain low-crystallinity apatites in the laboratory, with a particle size no greater than 50 nm, this is not the proper route; it is necessary to rely on wet-route methods, where the precursor salts are in solution, allowing a more homogeneous distribution of the components, almost at the atomic scale, where no kinetic impediments restrict the contact between reactants and give rise to a final product that is obtained faster and with less energy.

2.2.2 Synthesis of Apatites by Wet Route Methods

Synthetic apatites aimed at emulating the biological scenario should exhibit small particle sizes and the presence of CO_3^{2-}. In this sense, the wet route is the most suitable method of synthesis. There are several methods leading to nanometric-size apatites.

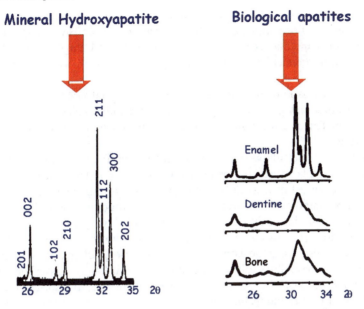

Figure 2.6 XRD diffraction patterns evidence the different crystallinity of mineral apatites *vs.* biological ones.

The alternative to enable the reactivity of solids and alleviate the problems of their diffusion is in the *wet route methods*. Working with reactants in solution, diffusion is now a simple phenomenon that enables chemical reactions at much lower temperatures. Besides, this method based on solutions not only simplifies the synthesis of the final product, it also achieves higher-quality products with a more homogeneous distribution of its components, higher reactivity, a decrease in reaction temperatures and heating periods, higher density of the final product and a smaller particle size. All this is related to the homogeneity of the starting materials and the use of lower synthesis temperatures.

Wet route methods modify the first stages of the reaction and allow a more efficient and complete *solid–solid* reaction at the last stage, with an easier diffusion process. As a consequence, many properties of the obtained solids are significantly modified and improved with these procedures.

Several methods have been tested trying to improve the homogeneity both in composition and particle size. These solution techniques can be classified in two large categories: *coprecipitation* and *sol-gel*. Both types allow solids to be obtained without previous milling and in a single calcination stage.

2.2.2.1 Sol-Gel

The chemistry of the sol-gel process is based on the hydrolysis and condensation of molecular precursors. There are two different routes described in the literature, depending on whether the precursor is formed by an aqueous solution of an inorganic salt or by a metal-organic compound. In any case, this method requires the careful study of parameters such as oxidation states, pH and concentration.

The sol-gel method is a process divided in several stages where different physical and chemical phenomena are performed, such as *hydrolysis, polymerisation, drying* and *densification*.

This process is known as *sol-gel* because differential viscosity increases at a given instant during the process sequence. A sudden increase in viscosity is a common feature in all *sol-gel* processes, which indicates the onset of *gel* formation.

Sol-gel processes allow synthesis of oxides from inorganic or metal-organic precursors, the latter usually being metal alkoxides. A large part of the literature on the sol-gel process deals with synthesis from alkoxides.

The most important features of the *sol-gel* method are: better homogeneity, compared with the traditional ceramic method, high purity of the obtained products, low processing temperatures, very uniform distribution in multi-component systems, good control of size and morphology, allows new crystalline or noncrystalline solids to be obtained and, finally, easy production of thin films and coatings. As a consequence of all this, the *sol-gel* method is widely used in ceramic technology.

The six main stages in *sol-gel* synthesis are depicted in Figure 2.7 and are defined as follows:

- *Hydrolysis*: The hydrolysis process may start with a mixture of metal alkoxides and water in a solvent (usually alcohol) at ambient or slightly

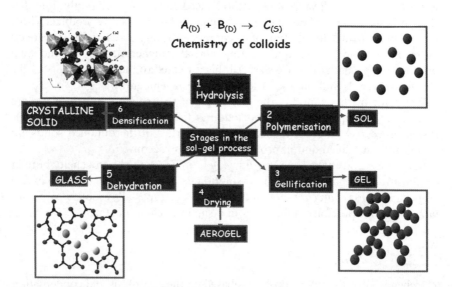

Figure 2.7   The sol-gel stages that allow sols, gels, aerogels, glasses or crystalline solids to be obtained.

higher temperature. An acid or basic catalyst may be added to increase the reaction rate.

- *Polymerisation*: At this stage, neighbouring molecules are condensed, water and alcohol are removed from them and the metal–oxide bonds are formed. The polymer network grows to colloidal dimensions in the liquid state (*sol*).
- *Gelification*: In this stage, the polymer network keeps growing until a three-dimensional network is formed through the ligand. The system becomes slightly stiff, which is a typical feature of a *gel* upon removal of the *sol* solvent. The solvent, water and alcohol molecules remain, however, inside the *gel* pores. The addition of smaller polymer units to the main network continues progressively with *gel* ageing.
- *Drying*: Water and alcohol are removed at mild temperatures (<470 K), giving rise to hydroxylated metal oxides with a residual organic content. If the aim is to prepare an *aerogel* with high specific surface and low density, the solvent must be removed under supercritical conditions.
- *Dehydration*: This stage is performed between 670 and 1070 K to remove the organic residue and chemically bonded water. The result is a metal oxide in glass or microcrystalline form, with microporosity higher than 20–30%.
- *Densification*: At temperatures above 1270 K we can obtain dense materials, due to the reaction between the various components of the precursor in the previous stage.

These six stages may or may not be followed strictly in practice; this will depend on the solid to be synthesised. As Figure 2.7 illustrates, the choice will be different if the purpose is obtaining an aerosol, a glass or a crystalline material.

It is possible to prepare crystalline materials following modifications of the described sol-gel route, without the addition of metal alkoxides. For instance, a solution of transition-metal salts can be transformed to a gel by adding an adequate organic reactant (*e.g.* 2-ethyl-1-hexanol). Alumina gels are prepared by ageing salts obtained by hydrolysis of aluminium butoxide followed by hydrolysis in hot water and peptisation with nitric acid.

Sol-gel process for HA (hydroxyapatite) synthesis usually can produce fine-grain microstructure containing a mixture of nano-to-submicrometre crystals.[69] Low-temperature formation and fusion of the apatite crystals have been the main contributions of the sol-gel process in comparison with conventional methods for HA powder synthesis. A number of combinations between calcium and phosphorus precursors were employed for sol-gel HA synthesis. For instance Liu *et al.*[70] used a triethyl phosphate sol that was diluted in anhydrous ethanol and then a small amount of distilled water was added for hydrolysis. The molar ratio of water to the phosphorus precursor is kept at 3. The mixture is then sealed and stirred vigorously. After approximately 30 min of mixing the emulsion transforms into a clear solution suggesting that the phosphate was completely hydrolysed. A stoichiometric amount of calcium nitrate is subsequently dissolved in anhydrous ethanol, and dropped into the hydrolysed phosphorus sol. As a result of this process, a clear solution was obtained and aged at room temperature for 16 h before drying. Further drying of the viscous liquid at temperatures about 60 °C results in a white gel, which can be treated at temperatures ranging between 600 to 1100 °C as a function of the particle size desired.

The major limitation of the sol-gel technique application is linked to the possible hydrolysis of phosphates and the high cost of the raw materials. Recently, Fathi and Hanifi[71] have developed a new sol-gel strategy to tackle this problem by using phosphoric pentoxide and calcium nitrate tetrahydrate. This sol-gel method provides a simple route for synthesis of hydroxyapatite nanopowder, where the crystalline degree and morphology of the obtained nanopowder are also dependent on the sintering temperature and time.

2.2.2.2 Solidification of Liquid Solutions

There are many variations and modifications to the sol-gel method. For instance, both simple and mixed oxides can also be synthesised by decomposition of metal salts of polybasic carboxylic acids, such as citrates. This procedure, however, can only be followed when the metal cations are soluble in organic solvents.

Solidification of this solution can be achieved by addition of a diol, for instance, which greatly increases the viscosity of the solution, due to the formation of three-dimensional ester-type polymers. When the diol reacts with the citric solution, a resin is formed that avoids the partial segregation of the components, preserving the homogeneity of the solution that is now in solid form.

The organic matter is removed by calcination at temperatures above 450 °C. A subsequent thermal treatment of the residue enables the solid–solid reaction in an easy and complete way, at temperatures lower than those needed for the ceramic method, as a consequence of the small particle size and the good homogeneity of all components in the matrix. This method was developed by Pechini and is known as the liquid solutions solidification technique (LSST).[72] Figure 2.8 depicts the different stages of this method.

The application of the liquid mix technique is based on the Pechini patent.[72] This patent was originally developed for the preparation of multicomponent oxides, allowing the production of massive and reproducible quantities with a precise homogeneity in both composition and particle size. This method is based on the preparation of a liquid solution that retains its homogeneity in the solid state. This method not only allows a precise control of the cation concentration, but also the diffusion process is enormously favoured by means of the liquid solution, compared to other classical methods. Its application has now extended to the preparation of calcium phosphates. The main difficulty of this synthesis lies on the presence of PO_4^{3-} groups that cannot be complexed by citric acid, and may cause its segregation and the formation of separated phosphate phases. The success of this task would suggest the possibility, by

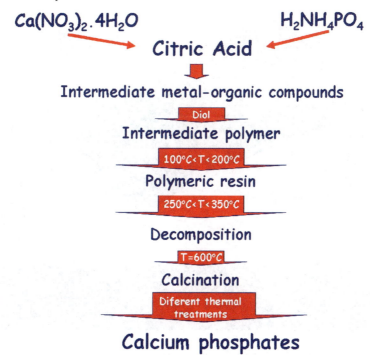

Figure 2.8 LSST application aims to obtain calcium phosphates.

modifying the synthesis conditions, of obtaining large amounts of single phases or biphasic mixtures with precise proportions of the calcium phosphates. This method makes it possible to obtain single-phase hydroxyapatite, β-TCP and α-TCP and also biphasic materials whose content in β-TCP and HA can be precisely predicted from the Ca/P ratio in the precursor solutions (Figure 2.9).[73]

2.2.2.3 Controlled Crystallisation Method

Methods based on precipitation from aqueous solutions are most suitable for preparation of large amounts of apatite, as needed for processing both into ceramic bodies and in association with different matrices. The difficulty with most of the conventional precipitation methods used is the synthesis of well-defined and reproducible orthophosphates.[8,9]

Problems can arise due to the usual lack of precise control on the factors governing the precipitation, pH, temperature, Ca/P ratio of reagents, *etc.*, which can lead to products with slight differences in stoichiometry, crystallinity, morphology, *etc.*, that could then contribute to the different "*in vivo/in vitro*" behaviours described. In this sense, it is important to develop a methodology able to produce massive and reproducible quantities of apatite, optimised for any specific application or processing requirements by controlling composition, impurities, morphology, and crystal and particle size. For quantitative reactions in

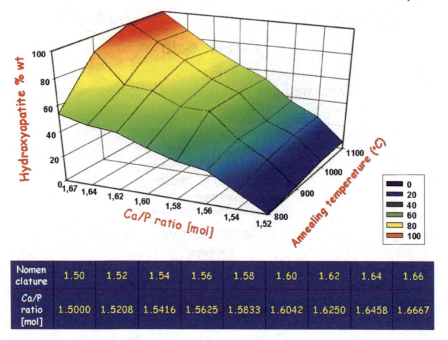

Nomen clature	1.50	1.52	1.54	1.56	1.58	1.60	1.62	1.64	1.66
Ca/P ratio [mol]	1.5000	1.5208	1.5416	1.5625	1.5833	1.6042	1.6250	1.6458	1.6667

Figure 2.9 Percentage of hydroxyapatite as a function of Ca/P ratio and annealing temperature when liquid mix technique is applied.

solutions, the reactants must be calcium and phosphate salts with ions that are unlikely to be incorporated into the apatite lattice. Since it has been claimed that NO_4^- and NH_4^+ are not incorporated into crystalline apatites, or in the case of NH_4^+ present a very limited incorporation,[74] the chosen reaction for this method was $10Ca(NO_3)_2 \cdot 4H_2O + 6(NH_4)_2HPO_4 + 8NH_4OH \rightarrow Ca_{10}(PO_4)_6(OH)_2 + 20NH_4NO_3 + 6H_2O$.

Thus, apatites with different stoichiometry and morphology can be prepared and the effects of varying synthesis conditions on stoichiometry, crystallinity, and morphology of the powder can be analysed. The effects of varying concentration of the reagents, the temperature of the reaction, reaction time, initial pH, ageing time, and the atmosphere within the reaction vessel can also be controlled with equipment like that represented in Figure 2.10. Temperatures in the range of 25–37 °C are necessary to obtain apatites with crystal sizes in the range of adult human bone, while 90 °C is necessary to obtain apatites with crystal sizes in the range of enamel. Higher reaction times lead to apatites with higher Ca/P ratios. Ageing of the precipitated powder can lead to the incorporation of minor quantities of carbonate. It is possible to force the incorporation of carbonate ions into the apatite structure without introducing monovalent cations.[5] The main results of the studied variations in the reaction conditions are, in short, that higher concentrations of reagents produce higher amounts of products with minor differences in their characteristics, allowing the production of homogeneous sets of materials.

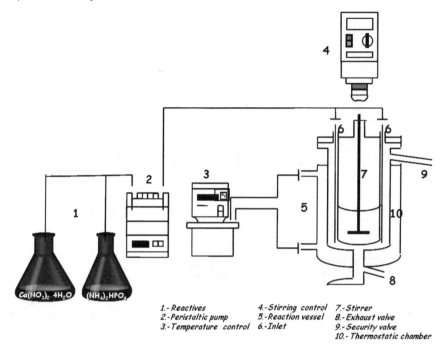

1.- Reactives
2.- Peristaltic pump
3.- Temperature control
4.- Stirring control
5.- Reaction vessel
6.- Inlet
7.- Stirrer
8.- Exhaust valve
9.- Security valve
10.- Thermostatic chamber

Figure 2.10 Scheme of the equipment used for the apatite synthesis through the controlled crystallisation method.

2.2.3 Synthesis of Apatites by Aerosol Processes

Aerosol-based processes can be considered as a type of solid synthesis that involves the transformation from *gas* to *particle* or from *droplet* to *particle*. When the aerosol reaches the reaction area, different decomposition phenomena may take place, depending on the precursor features and the temperature, as shown in Figure 2.11.

At low deposition temperatures, the droplets reach the substrate in liquid form. The solvent evaporates leaving a finely divided precipitate onto the substrate (layout A).

At higher temperatures, the solvent may evaporate before coming into contact with the substrate and the precipitate impacts the substrate (layout B).

When the deposition temperature is high enough and the precursor is volatile, there is a consecutive solvent evaporation and solute sublimation. This solute, in vapour form, diffuses to the substrate where a heterogeneous chemical reaction with its surface is performed, in the solid state. This is the so-called chemical deposition technique in vapour phase, or chemical vapour deposition (CVD, layout C).

At high temperatures, the reaction is verified before the vapours reach the substrate; hence it is a homogeneous reaction. The product of this reaction is deposited onto the substrate as a finely divided powder (layout D).

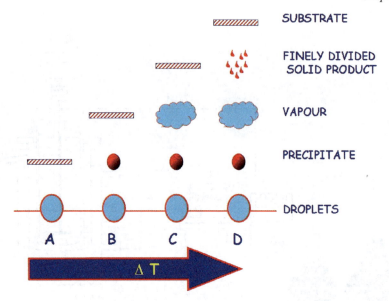

Figure 2.11　Different decomposition phenomena taking place depending on the temperature in an aerosol-assisted process.

In a gas–particle transformation, gases or vapours react forming primary particles that then start to grow by coagulation or by surface reactions. The powdered solids obtained with this process exhibit a narrow range distribution in sizes, and the method may yield spherical nonporous particles.

In the droplet to particle transformation processes, droplets containing the solute are suspended in a gaseous medium through a liquid atomisation where the droplets react with gases or are pyrolysed at high temperatures to form powder solids. The particle-size distribution is determined by the droplet size or by the processing conditions. The most frequent industrial methods to obtain powder solids from droplet to solid transformations are *drying* from an aerosol or *pyrolysis* from an aerosol. *Droplet freeze drying* is another technique in which powder solids are obtained by particle formation from droplets (Figure 2.12).

There are many different methods available to prepare ceramic thin films. Each method exhibits unique features that play a crucial role in the properties of the obtained particles. Therefore, it is important to choose wisely the deposition technique, and the working conditions, such as temperature, pressure, atmosphere or starting reactants.

Aerosol synthesis technique has been used to produce small particles of different materials.[75,76] Its main advantage is that this technique has the potential to create particles of unique composition, for which starting materials are mixed in a solution at atomic level (Figure 2.12). A better thermal treatment can originate important modifications on morphology and texture. In consequence, HA preparation by this method was deemed of interest.[77] Hydroxyapatite hollow particles have been prepared by pyrolysis of an aerosol produced by

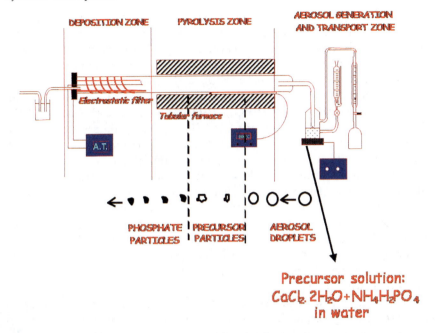

Figure 2.12 Scheme of the equipment used for the synthesis of powder solids from aerosol droplets.

ultrahigh frequency spraying of a $CaCl_2$-$(NH_4)H_2PO_4$ solution. Hollow particles were annealed at different temperatures. Thermal treatment at $1050\,°C$ produces the growth of nucleated crystallites in the particle surface, with remarkable morphology. The particle size range is 0.3–$2.2\,\mu m$. Apatite nanocrystals grow onto this surface.

2.2.4 Other Methods Based on Precipitation from Aqueous Solutions

2.2.4.1 Calcium Phosphate Cements as Apatite Precursors

Cements based on calcium salts, phosphates or sulfates, have attracted much attention in medicine and dentistry due to their excellent biocompatibility and bone-repair properties.[78–81] Moreover, they have the advantage over the bioceramics that they do not need to be delivered in prefabricated forms, because these self-setting cements can be handled by the clinician in paste form and injected into bone cavities. Depending on the cement formulation, or the presence of additives, different properties, such as setting time, porosity or mechanical behaviour have been found in these materials.[82–86]

On the other hand, in the literature on phosphates focused on calcium phosphate cements, the technique employed for obtaining such cements is to mix the different components; one of them is responsible for curing the mixture. For instance, in the Constanz cement[87] – the first of its kind to be commercialised – the final product is a carbonate-apatite (dahllite) with low crystallinity and a carbonate content reaching 4.6%, in substitution of phosphate groups (B-type carbonateapatite) as is the case in bones. Constanz cement is obtained from a dry mixture of α-tricalcium phosphate, α-$Ca_3(PO_4)_2$, calcium phosphate monohydrate, $Ca(H_2PO_4) \cdot H_2O$ and calcium carbonate, $CaCO_3$. The Ca/P ratio of the first component is 1.50, and 0.5 for the second one, both values significantly lower than the Ca/P ratio of 1.67 for hydroxyapatite. A liquid component – a sodium monoacid phosphate solution – is then added to this solid mixture, which allows the formation of an easily injectable paste that will cure over time. The paste curing happens after a very reasonable period of time when considering its use in surgery. In fact, after five minutes it shows a consistency suitable for injection, and upon ten minutes it is solid without any exothermal response, exhibiting an initial strength of 10 MPa. 12 h later, 90% of its weight has evolved to dahllite, with compression strength of 55 MPa, and 2.1 MPa when under stress. This cement is then resorbed and gradually replaced by newly formed bone.

Calcium phosphate cements that can be resorbed and injected are being commercialised by various international corporations,[88] with slight differences in their compositions and/or preparation. Research is still under way in order to improve the deficiencies still present.

These cements are very compatible with the bone and seem to resorb slowly; during this gradual process, the newly formed bone grows and replaces the cement. However, the properties of the calcium phosphate cements are still insufficient for their reliable application. There are problems related to their mechanical toughness, the curing time, the application technique on the osseous defect and the final biological properties. New improvements in the development of these cements will soon be described, solving at least in part some of these disadvantages. For instance, the curing time will be shortened, even in contact with blood, and the toughness under compression will also improve.

Most of the injectable calcium phosphate cements used evolve to an apatitic calcium phosphate during the setting reaction. One of the main drawbacks of these apatitic cements is the slow resorption rate of the apatite. On the other hand, calcium sulfate dihydrate, gypsum, has been used as bone-void filler during many years,[78,87–89] although it presents too fast a resorption rate to provide a good support for new bone. The combination of both, calcium sulfate and apatite, can overcome the individual drawbacks, and in recent years studies using this biphasic material have been performed.[90–92]

Despite the advantages, all these implants can act as foreign bodies and become potential sources of infections. Then, the *in vivo* use of these materials requires a preventive therapy and this may be achieved by introducing a drug into them, which can be locally released "*in situ*" after implantation. In fact, different studies using bioceramics and self-setting materials containing active drugs have been performed in recent years.[93–97]

In this sense, the addition of an antibiotic to calcium sulfate-based cements has also been studied, in order to determine if the presence of the drug affects the physical-chemical behaviour of the cements and to study the release kinetics of the drug from the cement. Two system types were chosen: gypsum and apatite/gypsum. The antibiotic chosen for this study was cephalexin in crystalline form, *i.e.* cephalexin monohydrate.

The presence of cephalexin into the cements does not alter neither the physical-chemical behaviour of the cements nor produce structural changes in them. The release of the drug is different depending on the composition. For gypsum cements, the cephalexin is quickly released, helped by a dissolution process of the matrix, whereas the drug release is more controlled by the hydroxyapatite presence in hydroxyapatite/gypsum samples. Apatite-containing cements do not only show a different drug-release process, also the paste viscosity is lower and a faster formation "*in vitro*" of an apatite-type layer on their surface is observed.[98]

2.2.4.2 Biphasic Mixtures of Calcium Phosphates as Apatite Precursors

Several attempts have been made to synthesise the mineral component of bones starting from biphasic mixtures of calcium phosphates.[99] Hence, bone-replacing materials based on mixtures of hydroxyapatite and β-TCP have been prepared; under physiological conditions, such mixtures evolve to carbonate hydroxyapatite. The chemical reactions are based in equilibrium conditions between the more stable phase, hydroxyapatite, and the phase prone to resorption, β-TCP. As a consequence, the mixture is gradually dissolved in the human body, acting as a stem for newly formed bone and releasing Ca^{2+} and PO_4^{3-} to the local environment. This material can be injected, used as coating or in any other form suitable for application as bulk bone replacement – forming of bulk pieces, filling of bone defects.[100] At present, a wide range of biphasic mixtures are under preparation, using various calcium phosphates, bioactive glasses, calcium sulfates, *etc.*[92,101,102]

Currently, there is an increasing interest in the preparation of mixtures of two or more calcium phosphates. These materials are commonly prepared with hydroxyapatite and a more resorbable material such as tricalcium phosphate (α or β) or calcium carbonate in different proportions depending on the characteristics required for a specific application. Some examples of commercial products based on these mixtures are: Triosite™, MBCPTM™, Eurocer®, *etc.* The synthesis routes commonly employed in the preparation of these mixtures include the blending of different calcium phosphates,[103,104] and precipitation.[105–107] Other techniques also employed are: solid state,[108] treatment of natural bone,[109] spray pyrolysis,[110] microwave,[111] combustion,[112] *etc.* Some authors have defended the superior properties of the biphasic materials "directly" prepared over those obtained by mixing two single phases.[113]

The promising results obtained with *cements* and *biphasic mixtures* seem to indicate that it is easier to obtain precursors of synthetic apatites that, when in contact with the biological environment, can evolve towards similar compositions to that of the biological apatite, than to obtain apatites in the laboratory with similar compositional and structural characteristics to those of the biological material, and in adequate quantities, *i.e.* large, industry-scale amounts with precise composition and easily repeatable batch after batch, for its use in the production of ceramic biomaterials.

Bioceramics aimed at the replacement or filling of bones could be obtained by synthesis of apatite precursors through different calcium phosphate mixtures, using a wet route. If the information gathered from the calcium cements is put to use, it would be necessary to eliminate the solution added to cure the mixture and search for compositions and ratios that allow to obtain precursors that, when in contact with the body fluids, evolve chemically towards the formation of carbonate hydroxyapatite crystals, with small particle size and low crystallinity, calcium deficient and with a carbonate content of approximately 4.5% w/w, located in the PO_4^{3-} sublattice.

2.2.5 Apatites in the Absence of Gravity

Particular attention must be paid at this point to the essays performed in the absence of gravity. Suvurova and Buffat[114] have compared the results obtained when calcium phosphate specimens, in particular, HA and triclinic octacalcium phosphate (OCP), are prepared from aqueous solutions under different conditions of precipitation. When supersaturated solutions of calcium phosphates are prepared by diffusion-controlled mixing in outer space (EURECA 1992–1993 flight) several differences are observed in crystal size, morphology and structural features with respect to those prepared on earth. It is worth stressing that space-grown OCP crystals possess a maximum growth rate in the [001] direction and a minimum rate in the [100] one. Space-grown and terrestrial HA crystals differ from each other in size: the former are at least 1–1.5 orders of magnitude bigger in length. Diffusion-controlled mixing in space seems to provide a lower supersaturation in the crystallisation system compared to earth, promoting the crystal growth in the competition between nucleation and growth. These authors conclude that similar processes may most probably arise in the human body (under definite internal conditions) during space flying when quite large HA crystals start to grow instead of the small and natural ones. In addition, other modifications of OCP crystals with huge sizes appear. These elements may disturb the Ca dynamical equilibrium in the body, which might lead to possible demineralisation of the bone tissue.

2.2.6 Carbonate Apatites

Biological apatites (mineral component of the bones) are difficult to synthesise in the laboratory with carbonate contents equivalent to those in the bone. Although the carbonate inclusion in itself is very simple[115] (in fact, when producing stoichiometric apatites in the laboratory, a strict control of the synthesis conditions is needed to avoid carbonate inclusion), the carbonate content is always different from the fraction of carbonates in the natural bone $(4-8 \, wt\%)^6$ and/or are located in different lattice positions.[116] At this point, it should be mentioned that this carbonate content can be slightly different when analysed samples come from other vertebrates.[117] The carbonate easily enters into the apatite structure, but the problem lies in the amount that should be introduced taking into account the carbonate content of biological apatites. When the aim is to obtain carbonate hydroxyapatite and the reaction takes place at high temperatures, the carbonates enter and occupy lattice positions in the OH^- sublattice (A-type apatites). In contrast, the carbonates in biological apatites always occupy positions in the PO_4^{3-} sublattice (that is, they are B-type apatites).[118] In order to solve this problem, low-temperature synthesis routes have to be followed, allowing carbonate hydroxyapatites to be obtained with carbonates in phosphate positions.[6] But the amount entered remains to be solved, and it is usually lower than the carbonate content of the mineral component of the bones.

These calcium-deficient and carbonated apatites have been obtained in the laboratory by various techniques; nowadays, it is known that apatites with low crystallinity, calcium deficiency and carbonate content can be obtained, but with carbonate contents usually unequal to those of the natural bones.[5,119,120] Therefore, the main problem remains in the control of carbonate content and lattice positioning.

2.2.7 Silica as a Component in Apatite Precursor Ceramic Materials

One way to enhance the bioactive behaviour of hydroxyapatite is to obtain substituted apatites, which resemble the chemical composition and structure of the mineral phase in bones.[121,122] These ionic substitutions can modify the surface structure and electrical charge of hydroxyapatite, with potential influence on the material in biological environments. In this sense, an interesting way to improve the bioactivity of hydroxyapatite is the addition of silicon to the apatite structure, taking into account the influence of this element on the bioactivity of bioactive glasses and glass-ceramics.[123,124] In addition, several studies have proposed the remarkable importance of silicon on bone formation and growth[125,126] under *in vitro* and *in vivo* conditions.

Several methods for the synthesis of silicon-substituted hydroxyapatites have been described. Ruys[127] suggested the use of a sol-gel procedure; however, these materials, besides the hydroxyapatite phase, include other crystalline phases depending on the substitution degree of silicon. Tanizawa and Suzuki[128] tried hydrothermal methods, obtaining materials with a $Ca/(P + Si)$ ratio higher than that of pure calcium hydroxyapatite. Boyer *et al.*[129] conducted studies on the synthesis of silicon-substituted hydroxyapatites by solid-state reaction, but in these cases the incorporation of a secondary ion, such as La^{3+} or SO_4^{2-}, was needed. In these examples, no bioactivity studies were performed on the silicon-containing apatites.

Gibson *et al.*[130] synthesised silicon-containing hydroxyapatite by using a wet method, and its *in vitro* bioactivity studies gave good results. These authors studied the effects of low substitution levels on the biocompatibility and *in vitro* bioactivity, determining the ability to form the apatite-like layer by soaking the materials in a simulated body fluid (SBF).[131] Also, Marques *et al.*[132] synthesised, by wet method, hydroxyapatite with silicon content up to 0.15 wt%, obtaining stable materials at 1300 °C and noting that the unit cell volume and the *a* parameter length of the hydroxyapatite decreased as the silicon content increased.

Hence, the role of silicon substituting part of the phosphorus atoms present in the hydroxyapatite lattice seems to be an important factor influencing the bioactive behaviour of the material. However, it is not clearly known whether the silicon present in the material substitutes completely the phosphorus in the hydroxyapatite structure, or whether the replacement is partial, or even if in any of the described synthesis the silicon species remain as an independent phase. In all the cited syntheses, the final product contains silicon, but its chemical nature is not revealed.

A similar work focused on the synthesis and bioactivity study of hydroxyapatites containing orthosilicate anions that isomorphically replace phosphate groups, aimed at improving the bioactivity of the resulting materials as compared with that of pure calcium hydroxyapatite.[133] To accomplish this purpose, two synthesis procedures were used, starting from two different calcium and phosphorus precursors and the same silicon reagent in both cases. To assess the proposed substitution, surface chemical and structural characterisation of the silicon-substituted hydroxyapatites was performed by means of X-ray diffraction (XRD) and X-ray photoelectron spectroscopy (XPS). The *in vitro* bioactivity of the so-obtained materials was determined by soaking the materials in SBF and monitoring the changes of pH and chemical composition of the solution, whereas the modification at the surface was followed by means of XPS, XRD, and scanning electron microscopy (SEM). Silicon-containing hydroxyapatites were synthesised by the controlled crystallisation method. Chemical analysis, N_2 adsorption, Hg porosimetry, X-ray diffraction, scanning electron microscopy, energy-dispersive X-ray spectroscopy, and X-ray photoelectron spectroscopy were used to characterise the hydroxyapatite and to monitor the development of a calcium phosphate layer onto the substrate surface immersed in a simulated body fluid, that is, *in vitro* bioactivity tests. The influence of the silicon content and the nature of the starting calcium and phosphorus sources on the *in vitro* bioactivity of the resulting materials were studied. A sample of silicocarnotite, whose structure is related to that of hydroxyapatite and contains isolated SiO_4^{4-} anions that isomorphically substitute some PO_4^{3-} anions, was prepared and used as reference material for XPS studies. An increase of the unit cell parameters with the Si content was observed, which indicated that SiO_4^{4-} units are present in lattice positions, replacing some PO_4^{3-} groups. By using XPS it was possible to assess the presence of monomeric SiO_4^{4-} units in the surface of apatite samples containing 0.8 wt% of silicon, regardless of the nature of the starting raw materials, either $Ca(NO_3)_2$/ $(NH_4)_2HPO_4$/Si-$(OCOCH_3)_4$ or $Ca(OH)_2$/H_3PO_4/Si$(OCOCH_3)_4$. However, an

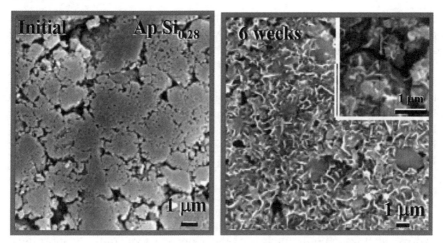

Figure 2.13 Scanning electron micrographs of silicon-substituted apatites before and after soaking for 6 weeks in simulated body fluid.

increase of the silicon content up to 1.6 wt% leads to the polymerisation of the silicate species at the surface. This technique shows silicon enrichment at the surface of the three samples. The *in vitro* bioactivity assays showed that the formation of an apatite-like layer onto the surface of silicon-containing substrates is strongly enhanced as compared with pure silicon-free hydroxyapatite (Figure 2.13). The samples containing monomeric silicate species showed higher *in vitro* bioactivity than that of a silicon-rich sample containing polymeric silicate species. The use of calcium and phosphate salts as precursors leads to materials with higher bioactivity.[133]

Finally, the results revealed that controlled crystallisation is a good procedure to prepare silicon-substituted hydroxyapatites that can be used as a potential material for prosthetic applications.

The presence of silicon (Si) in HA has shown an important role on the formation of bone.[122] To study the role of Si, Si-substituted hydroxyapatite (SiHA) has been synthesised by several methods[127–130,132,133] but its structural characteristics and microstructure remain not fully understood. Most of the structural studies carried out until now (mainly by X-ray diffraction) had not demonstrated the Si incorporation into the apatite structure. In fact, the very similar scattering factor makes it very difficult to determine if Si has replaced some P in the same crystallographic position. The absence of secondary phases and the different bioactive behaviour were the best evidence for Si incorporation. No positive evidence or quantitative study of P substitution by Si has been carried out yet. On the other hand, the hydroxyl groups sited at the *4e* position are one of the most important sites for the HA reactivity. The movement of H along the *c*-axis contributes to the HA reactivity. However, XRD is not the optimum tool for the study of light atoms such as H; neutron diffraction (ND) is an excellent alternative to solve this problem. The Fermi lengths for Si and P are different enough to be discriminated, whereas neutrons are very sensitive to the H presence. Consequently, XRD and ND have combined to answer one of the most recent questions in the dentistry and orthopedic surgery fields.

In order to explain the higher bioactivity of the silicon-substituted hydroxyapatite, synthetic ceramic hydroxyapatite (HA) and SiHA have been structurally studied by neutron scattering. The Rietveld refinements show that the final compounds are oxy-hydroxyapatites, when obtained by solid-state synthesis under an air atmosphere. By using neutron diffraction, the substitution of P by Si into the apatite structure has been corroborated in these compounds. Moreover, these studies also allow us to explain the superior bioactive behaviour of SiHA, in terms of higher thermal displacement parameters of the H located at the *4e* site.[134]

Structure refinements by the Rietveld method indicate that the thermal treatment produces partial decomposition of the OH groups, leading to oxy-hydroxy apatites in both samples and the higher reactivity of the Si-substituted HA can be explained in terms of an increasing of the thermal ellipsoid dimension parallel to the *c*-axis for H atoms (Figure 2.14).

2.2.8 Apatite Coatings

The application of synthesis methods to obtain apatite coatings is a subject of great interest nowadays in the field of biomaterials, in relation to the fabrication of load-bearing implants made of metal alloys.[135] When said metal implants are coated with a ceramic material such as apatite, the performance of the implant

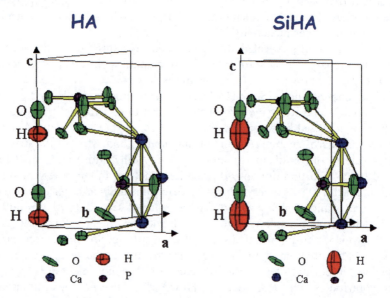

Figure 2.14   Thermal ellipsoids of HA (left) and SiHA (right). The thermal ellipsoids for H atoms are more than twice as big for SiHA than for HA.

improves due to the barrier effect of the apatite coating against metal ion diffusion from the implant towards the body, while enabling a better adhesion to the bone tissue. There are plenty of methods in use nowadays to fabricate this type of coatings. Some of the techniques used will be briefly described below, although the production of biomimetic coatings will be dealt with in detail in Chapter 5.

2.2.8.1 Production of Thin Films by Vapour-Phase Methods

Some of the vapour-phase methods in use are chemical vapour deposition (CVD), chemical transportation, substrate reaction, pulverisation pyrolysis, vacuum evaporation, sputtering, ion plating techniques and plasma pulverisation methods.

2.2.8.2 Production of Thin Films by Liquid-Phase Methods

Thin films can also be obtained from liquid precursors, using procedures such as sol-gel, where a gel is prepared with metal alkoxides or from organic or inorganic salts. The films are formed onto substrates by drying and heating a *sol* previously used to coat the substrates. There are different coating techniques available (Figure 2.15). This is a very popular coating process due to its ability to grow films onto substrates of very different shapes and sizes.

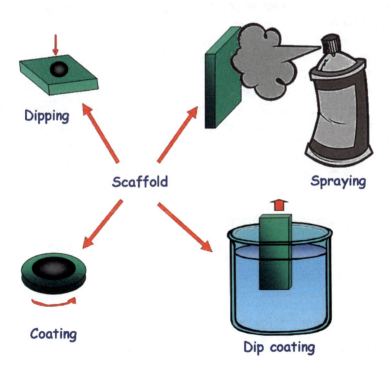

Figure 2.15 Different available coating techniques.

Besides sol-gel, other procedures in this category include *liquid phase epitaxy* and *melt epitaxy*.

2.2.8.3   Production of Thin Films by Solid-Phase Methods

Solids can also be used as precursors for thin films. For instance, the method of *thermal decomposition of a coating*, where the thin film is obtained by high-temperature decomposition of a metal-organic compound dissolved in an organic solvent that covers the substrate, or the *precipitation* method, where aqueous solutions of different salts react and the compounds with less solubility precipitate. Ceramic powder materials can be obtained after washing, drying and calcining the precipitates. Using different salts and controlling parameters such as the temperature, it is possible to control the particle size of the obtained solids, which can be synthesised at high temperatures to obtain polycrystalline films.

This method has been successfully applied to improve the biological osteo-blastic response,[136] and recently new silicon-substituted apatite coatings have been synthesised.[137,138] Basically, this last process is carried out by coating Ti substrates or, in the case of silicon-substituted calcium phosphates, on quartz substrates. The procedure can be briefly described as follows. Ammonium phosphate solution is titrated into an aqueous solution of calcium nitrate adding ammonium hydroxide to keep the pH at 10.5, according to the following reaction:

$$10Ca(NO_3)_2 \Downarrow\leftarrow 6NH_4H_2PO_4 \Downarrow\leftarrow 14NH_4OH \cdot \leftarrow Ca_{10}(PO_4)_6(OH)_2 \Downarrow$$
$$\leftarrow 20NH_4NO_3 \Downarrow 12H_2O$$

The resulting calcium phosphate solution is then aged at room temperature for 1 day and then concentrated. The corresponding substrate (Ti alloy, quartz, *etc.*) is dipped into the solution. After drying, the samples can be sintered at temperatures between 700 and 1100 °C depending on the final microstructure.

2.2.9   Precursors to Obtain Apatites

The different synthesis routes applied to obtain apatites require the use of precursors with certain features. Several potential precursors will be described and classified below:

2.2.9.1   Inorganic Salts

Inorganic salts are used as molecular precursors in wet route chemical processes, such as: *sol-gel*, *colloidal* or *hydrothermal*. They are ionic compounds, and some examples of these precursor salts are collected in Table 2.1.

Table 2.1 Some precursor salts.

Inorganic salts	Examples
Metal halides	$MgCl_2$, LiF, KCl, $SiCl_4$, $TiCl_4$, $CuCl_2$, KBr, $ZrOCl_2$
Metal carbonates	$MgCO_3$, $CaCO_3$, Na_2CO_3, $SrCO_3$
Metal sulfates	$MgSO_4$, $BaSO_4$, K_2SO_4, $PbSO_4$
Metal nitrates	$LiNO_3$, KNO_3, $Fe(NO_3)_2$
Metal hydroxides	$Ca(OH)_2$, $Mg(OH)_2$, $Al(OH)_3$, $Fe(OH)_3$, $Zr(OH)_4$
Salts with mixed ligands	$(CH_3)SnNO_3$, $(C_2H_5)_3SiCl$, $(CH_3)_2Si(OH)_2$

Table 2.2 Some coordination compound precursors.

Coordination compound	General formula	Selected examples
Metal alkoxides	-M(-OR), R is an alkyl	$Al(OC_3H_7)_3$, $Si(OCH_3)_4$, $Ti(OC_3H_7)_4$, $Zr(OC_4H_9)_4$
Metal carboxylates	$-M(-OC(O)R)_x$, R is an alkyl	$Al(OC(O)CH_3)_3$, $Pb(OC(O)CH_3)_2$ -acetates $Pb(OC(O)CH_2CH_3)_4$ -propionate $Al(OC(O)C_6H_5)_3$ -benzoate
Metal ketones	$-M(-OCRCH(R')CO)_x$, R is an alkyl or aryl	$Ca(OC(CH_3)CH(CH_3)CO)_2$ -pentanedionate $Al(OC(C(CH_3)_3)CH(C(CH_3)_3)$ $CO)_2$ -heptanedionate
Metal amines		$(CH_3)_2AlNH_2$, $(C_2H_5)_2AlN(CH_3)_2$, $(CH_3)BeN(CH_3)_2$, $(iC_3H_7)_3$- $GeNH_2$, $(C_3H_7)_3PbN(C_2H_5)_2$
Metal thiolates	$-M(-SR)_x$, R is an alkyl or aryl	$(CH_3)_2Ge(SC_2H_5)$, $Hg(C_4H_3S)_2$, $(SCH_3)Ti(C_5H_5)_2$, $(CH_3)Zn(SC_6H_5)$
Metal azides	$-MN_3$	$(CH_3)_3SnN_3$, CH_3HgN_3
Metal isothiocyanates	$-M(-NCS)_x$	$(C_2H_5)_3Sn(NCS)$
Coordination compounds with mixed functional groups		$(C_4H_9)Sn(OC(O)CH_3)_3$, $(C_5H_5)_2TiCl_2$, $(C_5H_5)Ti(OC(O)CH_3)_3$

2.2.9.2 Coordination Compounds with Organic Ligands

Coordination compounds with organic ligands are covalent or ionic coordination compounds in which the metal site is bonded to the ligand by an oxygen, sulfur, phosphorus or nitrogen atom. These compounds are used as precursors both in wet-route processes and in vapour-phase reactions. Table 2.2 shows some examples of coordination compounds with organic ligands.

2.2.9.3 Organometallic Compounds

Organometallic compounds are covalent or coordination compounds in which the ligand is bonded to the metal site by a carbon atom. As in the previous case

Table 2.3 Some organometallic precursors.

Organometallic compounds	Selected examples
Metal alkyls	$As(CH_3)_3$, $Ca(CH_3)_2$, $Sn(CH_3)_4$ -methyl
Metal aryls	$Ca(C_6H_5)_2$ -phenyl
Metal alkenyls	$Al(CH=CH_2)_3$ -vinyl,
	$Ca(CH=CHCH_3)_2$ -propenyl
Metal alkynyls	$Al(CCH)_3$, $Ca(CCH)_2$ -acetylene
Metal carbonyls	$Co_2(CO)_8$, $Mn_2(CO)_{12}$, $W(CO)_6$ -carbonyl
Mixed organic ligands	$Ca(CCC_6H_5)_2$ -phenylacetylene, $(C_5H_5)_3U(CCH)$
	-cyclopentadienyl/ethynyl

of coordination compounds with organic ligands, organometallic compounds are used as precursors in wet-route processes and in vapour-phase reactions. The most commonly used organometallic compounds are listed in Table 2.3.

2.2.9.4 Polymer Precursors

In some processes such as *metal-organic decomposition* and *sol-gel* processes, the polymer precursors can be used as starting materials to obtain glasses or ceramics. These polymers are often referred to as preceramic polymers. Table 2.4 shows some examples.

2.2.10 Additional Synthesis Methods

Different synthesis methods play a significant role in the design of apatites resembling their biological counterparts. Procedures related to the so-called *soft chemistry* are increasingly used and studied, since these methods allow new products to be obtained, many of them metastable and hence impossible to prepare by conventional routes such as the ceramic method. This "soft chemistry" applies to simple reactions that take place at relatively low temperatures, such as *intercalation*, *ionic exchange*, *hydrolysis*, *dehydration* and *reduction*. The advantage of using soft methods is that it is possible to better control the structure, stoichiometry and phase purity.

It is worth recalling that, opposed to this clear trend of avoidance of "extreme conditions" of synthesis, there is a method that actually opts for them. This is the *mechanochemical* method, where an intense and prolonged milling process is applied to generate, locally, high pressures and temperatures that lead to chemical transformations in the starting products, often obtaining metastable phases related to those special conditions of temperature and pressure.[139]

2.2.11 Sintered Apatites

In *sintering* processes, the particles are agglomerated. Sintering could be defined as a process where a compact solid changes its morphology and the size of its

Table 2.4 Some polymer precursors.

Polymer	Formula	
Polycarbosilanes	-[(RR')Si-CH$_2$-]$_x$	SiC precursor in *metal organic decomposition* and *sol-gel* processes, where R is an active functional group such as olefin, acetylene, H
Polysilazanes	-[(RR')Si-NR-]$_x$, R is an organic radical or H	Si$_3$N$_4$ or silicon carbonitride precursor, in a similar fashion to polycarbosilane
Polysiloxanes	1. -[Si(RR')O-]$_x$, linear R is an alkyl or aryl 2. silsesquioxanes: ladder structure 3. -[Si(CH$_3$)$_2$O-Si(CH$_3$)$_2$(C$_6$H$_5$)-]$_m$ 4. Random- or block-copolymers	Used in *sol-gel* processes and *in-situ* multiphase systems, as SiO$_2$ or silicon oxocarbide precursors
Polysilanes	-[Si(RR')-]$_n$, R is an alkyl or aryl	SiC precursors, for photo-resistants and photoinitiator materials
Borazines	-[BRNR'-]$_n$, cyclic units or in repeated chains	BN precursors in CVD/MOCVD and sol-gel processes.
Carboranes	B and C cage structures	B$_4$C precursors in MOCVD and MOD processes.
Polyphosphazenes	-[N=P(R$_2$)-]$_n$, R is an organic, organic metal or inorganic unit	Common substituents are: alkoxides, aryloxy, aryl-amides, carboxylates or halides
Polytinoxanes	-[Sn(R)$_2$-O-R'-O-Sn(R)$_2$-O]$_n$ chains, R is an organic group. Stair and drum structures are also possible.	
Polygermanes	-[Ge(RR')-]$_n$	Can be used in microlithographic applications as polysilanes

grains and *pores* through an atomic transport mechanism, so that the final result is a denser ceramic (Figure 2.16).

This transformation takes place when the powder material is subjected to a given pressure or to high temperature, or when placed under both effects simultaneously.

A sintering process may produce two different results:

1. A chemical transformation.
2. A simple geometrical rearrangement of the texture, defined here as the size and shape of the grains and pores in the solid. In this last case, the sintering process merely yields a product with identical chemical composition and crystalline structure to the initial material.

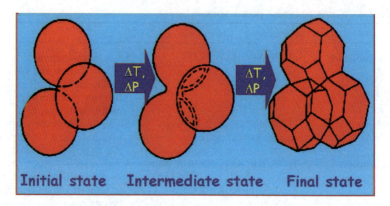

Figure 2.16   Stages of the sintering process.

However, sintering is more often used to improve the ceramic properties of a material, and to achieve higher values of packing and density. This occurs when a certain pressure is applied, without modifying the crystal structure, in order to modify some of its mechanical properties. The most common process includes a combination of pressure and temperature.

Regardless of the presence or not of structure variations, the sintering phenomenon implies a series of changes in the properties of materials, which can be summarised in the following points:

a) The agglomerate is contracted;
b) The pores change their shape and can even disappear;
c) The grain size increases;
d) The density increases.

Geometrical models help to understand the mechanisms involved in a sintering process with powder solids. A commonly used model considers equally sized spherical particles stacked together forming a compact packing and a three-stage sintering process:

Due to the heating effect, the particles are joined together during the first stage, and the empty spaces between them start to disappear or to decrease. In the second stage, *grain boundaries* start to form. And in the third stage, when the grain growth has been verified due to recrystallisation, the uniformly distributed pores are placed inside the grain and not in the boundaries. In this last stage, the larger pores grow at the expense of the smaller ones, due to their different chemical potential. Parameters such as *temperature, grain size, pressure* and *atmosphere* are very important in any sintering process. Figure 2.17 depicts the evolution of microstructure in a solid during a sintering process.

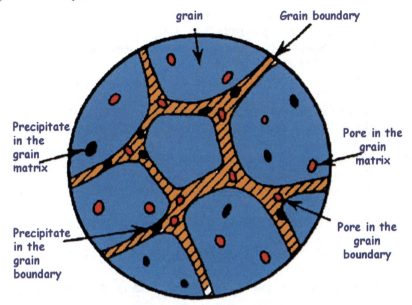

grain   Grain boundary

Precipitate in the grain matrix

Pore in the grain matrix

Precipitate in the grain boundary

Pore in the grain boundary

Figure 2.17 Evolution of microstructure in a solid during a sintering process.

These general remarks on sintering of solids are applicable to apatites, and are especially important when dealing with the fabrication of implants, since many of their features can be modified in this way, such as crystallinity, particle size and porosity.

References

1. C. N. R. Rao and J. Gopalakrishnan, *New Directions in Solid State Chemistry*, ed. Cambridge. University Press, United Kingdom, 1997.
2. M. Vallet-Regí, *Perspectives in Solid State Chemistry*, K. J. Rao ed., Narosa Publishing House, India, 1995, 37–65.
3. M. Vallet-Regí and J. González-Calbet, *Prog. Solid State Chem.*, 2004, **32**, 1.
4. M. Vallet-Regí, *J. Chem. Soc. Dalton Trans.*, 2001, 97.
5. L. M. Rodriguez-Lorenzo and M. Vallet-Regí, *Chem. Mater.*, 2000, **12**(8), 2460.
6. R. Z. LeGeros, In: *Monographs in Oral Science, Vol. 15: Calcium Phosphates in Oral Biology and Medicine*, H. M. Myers. ed. S. Karger, Basel, 1991.
7. J. C. Elliot eds., Sauramps Medical; Montpellier, 1998, 25–66.
8. W. Suchanek and M. Yoshimura, *J. Mater. Res.*, 1998, **13–1**, 94.
9. T. S. Narasaraju and D. E. Phebe, *J. Mater. Sci.*, 1996, **31**, 1.

10. K. de Groot, *Ceram. Int.*, 1993, **19**, 363.
11. M. Bohner, *J. Care Injured*, 2000, **31**, D37.
12. S. Sánchez-Salcedo, I. Izquierdo-Barba, D. Arcos and M. Vallet-Regí, *Tissue Eng.*, 2006, **12**(2), 279.
13. S. Padilla, S. Sánchez-Salcedo and M. Vallet-Regí, *J. Biomed. Mater. Res.*, 2005, **75A**, 63.
14. M. Vallet-Regí and D. Arcos, *J. Mater. Chem.*, 2005, **15**, 1509.
15. M. Vallet-Regí, J. Peña and I. Izquierdo-Barba, *Solid State Ion.*, 2004, **172**, 445.
16. D. Arcos, J. Rodríguez-Carvajal and M. Vallet-Regí, *Chem. Mater.*, 2004, **16**, 2300.
17. R. P. del Real, E. Ooms, J. G. G. Wolke and M. Vallet-Regí, *J.A. Jansen. J. Biomed. Mater. Res.*, 2003, **65A**, 30.
18. D. Arcos, R. P. del Real and M. Vallet-Regí, *J. Biomed. Mater. Res.*, 2003, **65A**, 71.
19. F. Balas, J. Pérez-Pariente and M. Vallet-Regí, *J. Biomed. Mater. Res.*, 2003, **66A**, 364.
20. S. Padilla, J. Román and M. Vallet-Regí, *J. Mater. Sci-Mater M.*, 2002, **13**, 1193.
21. C. V. Ragel, M. Vallet-Regí and L. M. Rodríguez-Lorenzo, *Biomaterials*, 2002, **23**, 1865.
22. M. V. Cabañas, L. M. Rodríguez-Lorenzo and M. Vallet-Regí, *Chem. Mater.*, 2002, **14**, 3550.
23. A. Rámila, S. Padilla, B. Muñoz and M. Vallet-Regí, *Chem. Mater.*, 2002, **14**, 2439.
24. L. M. Rodríguez, M. Vallet-Regí and J. M. F. Ferreira, *J. Biomed. Mater. Res.*, 2002, **60**, 232.
25. R. P. del Real, J. G. C. Wolke, M. Vallet-Regí and J. A. Jansen, *Biomaterials*, 2002, **23**, 3673.
26. L. M. Rodríguez-Lorenzo, M. Vallet-Regí and J. M. F. Ferreira, *Biomaterials*, 2001, **22**, 583.
27. L. M. Rodríguez, M. Vallet-Regí and J. M. F. Ferreira, *Biomaterials.*, 2001, **22**, 1847.
28. A. J. Salinas, M. Vallet-Regí and I. Izquierdo-Barba, *J. Sol-Gel Sci. Technol.*, 2001, **21**, 13.
29. M. Vallet-Regí and D. Arcos, *J. Pérez-Pariente. J. Biomed. Mater. Res.*, 2000, **51**, 23.
30. M. Vallet-Regí and A. Rámila, *Chem. Mater.*, 2000, **12**, 961.
31. M. Vallet-Regí, J. Pérez-Pariente, I. Izquierdo-Barba and A. J. Salinas, *Chem. Mater.*, 2000, **12**, 3770.
32. S. Sánchez-Salcedo, A. Nieto, and M. Vallet-Regí. *Chem. Eng.* J. DOI 10.1016/j.cej.2007.09011.
33. M. Vallet-Regí, I. Izquierdo-Barba and A. J. Salinas, *J. Biomed. Mater. Res.*, 1999, **46**, 560.
34. M. Vallet-Regí, A. M. Romero, V. Ragel and R. Z. Legeros, *J. Biomed. Mater. Res.*, 1999, **44**, 416.

35. I. Izquierdo-Barba, A. J. Salinas and M. Vallet-Regí, *J. Biomed. Mater. Res.*, 1999, **47**, 243.
36. M. Vallet-Regí, L. M. Rodríguez Lorenzo and A. J. Salinas, *Solid State Ion.*, 1997, **101–103**, 1279.
37. Tas. A. Cuneyt, *Biomaterials*, 2000, **21**, 1429.
38. Andrés-Vergés, C. Fernández-González, M. Martínez-Gallego, I. Solier, J. D. Cachadiña and E. Matijevic, *J. Mater. Res.*, 2000, **15–11**, 2526.
39. A. Yasukawa, T. Matsuura, M. Kakajima, K. Kandori and T. Ishikawa, *Mater. Res. Bull.*, 1999, **24**, 589.
40. J. Peña, I. Izquierdo-Barba, M. A. García and M. Vallet-Regí, *J. Eur. Ceram. Soc.*, 2006, **26**, 3631.
41. J. Peña, I. Izquierdo-Barba, A. Martínez and M. Vallet-Regí, *Solid State Sci.*, 2006, **8**, 513.
42. W. Weng and J. L. Baptista, *Biomaterials*, 1998, **19**, 125.
43. A. Jilavenkatesa and R. A. Condrate, *J. Mater. Sci.*, 1998, **33**, 4111.
44. C. S. Chai, K. A. Gross and B. Ben-Nissan, *Biomaterials*, 1998, **19**, 2291.
45. P. Layrolle, A. Ito and T. Tateishi, *J. Am. Ceram. Soc.*, 1998, **81–6**, 1421.
46. D. M. Liu, T. Trocynzki and W. J. Tseng, *Biomaterials*, 2001, **22**, 1721.
47. M. Manzano, D. Arcos, M. Rodríguez-Delgado, E. Ruíz, F. J. Gil and M. Vallet-Regí, *Chem Mater.*, 2006, **18**, 5696.
48. M. Vallet-Regí, *Dalton Trans.*, 2006, **1**, 5211–5220.
49. M. Vallet-Regí and D. Arcos, *Curr. Nanosci.*, 2006, **2**, 179.
50. J. Peña, I. Izquierdo-Barba and M. Vallet-Regí, *Key Eng. Mater.*, 2004, **254–256**, 359.
51. M. H. Fathi and A. Hanifi, *Mater. Lett.*, 2007, **61**, 3978.
52. T. S. Kumar Sampath, I. Manjubala and J. Gunasekaran, *Biomaterials*, 2000, **21**, 1623.
53. Y. Fang, D. K. Agrawal, D. M. Roy and R. Roy, *J. Mater. Res.*, 1992, **7**(2), 490.
54. K. Itatani, K. Iwafune, F. Scott Howellm and M. Aizawa, *Mater. Res. Bull.*, 2000, **35**, 575.
55. B. Yeong, J. M. Xue and J. Wang, *J. Am. Ceram. Soc.*, 2001, **84**, 465.
56. W. Kim, Q. Zang and F. Saito, *J. Mater. Sci.*, 2000, **35**, 5401.
57. B. Yeong, X. Junmin and J. Wang, *J. Am. Ceram. Soc.*, 2001, **82**, 65.
58. T. Nakano, A. Tokumura, Y. Umakoshi, S. Imazato, A. Ehara and S. Ebisu, *J. Mater. Sci.: Mater. Med.*, 2001, **12**, 703.
59. P. Shuk, W. L. Suchanek, T. Hao, E. Gulliver, R. E. Riman, M. Senna, K. S. TenHuisen and V. F. Janas, *J. Mater. Res.*, 2001, **16**, 1231.
60. G. K. Lim, J. Wang, S. C. Ng, C. H. Chew and L. M. Gan, *Biomaterials*, 1997, **18**, 1433.
61. D. Wals and S. Mann, *Chem. Mater.*, 1996, **8**, 1944.
62. T. Furuzono, D. Walsh, K. Sato, K. Sonoda and J. Tanaka, *J. Mater. Sci. Lett.*, 2001, **20**, 111.
63. S. Loher, W. J. Stark, M. Maciejewski, A. Baiker, S. E. Pratsinis, D. Reichardt, F. Maspero, F. Krumeich and D. Gunther, *Chem. Mater.*, 2005, **17**, 36.

64. M. Aizawa, T. Hanazawa, K. Itatani, F. S. Howell and A. Kishioka, *J. Mater. Sci.*, 1999, **34**, 2865.
65. D. Veilleux, N. Barthelemy, J. C. Trombe and M. Verelst, *J. Mater. Sci.*, 2001, **36**, 2245.
66. K. S. Tenhuisen and P. W. Brown, *Biomaterials.*, 1998, **19**, 2209.
67. W. Kim and F. Satio, *Ultrason. Sonochem.*, 2001, **8**, 85.
68. Y. Fang, D. K. Agrawal, D. M. Roy, R. Roy and P. W. Brown, *J. Mater. Res.*, 1992, **7**, 2294.
69. W. J. Weng and J. L. Baptista, *Biomaterials.*, 1998, **19**, 125.
70. D. M. Liu, T. Troczynski and W. J. Tseng, *Biomaterials.*, 2001, **22**, 1721.
71. M. H. Fathi and A. Hanifi, *Mater. Lett.*, 2007, **61**, 3978.
72. M. P. Pechini. (July 11, 1967) U. S. Patent 3,330,697; 1967.
73. J. Peña and M. Vallet-Regí, *J. Eur. Ceram. Soc.*, 2003, **23**(10), 1687–1696.
74. J. C. Elliott, *Studies in Inorganic Chemistry* 18. Elsevier, Amsterdam, 1994.
75. M. V. Cabañas, J. M. González-Calbet, M. Labeau, P. Mollard, M. Pernet and M. Vallet-Regí, *J. Solid State Chem.*, 1992, 101–265.
76. M. Vallet-Regí, V. Ragel, J. Román, J. L. Martínez, M. Labeau and J. M. González-Calbet, *J. Mater. Res.*, 1993, **8**(1), 138.
77. M. Vallet-Regí, M. T. Gutierrez-Ríos, M. P. Alonso, M. I. Frutos and S. Nicolopoulos, *J. Solid State Chem.*, 1994, **8**(1), 138.
78. A. S. Coetzee, *Arch. Otolaryngol.*, 1980, **106**, 405.
79. J. Lemaitre, A. Mirtchi and E. Munting, *Sil. Ind. Ceram. Sci. Technol.*, 1987, **52**, 141.
80. L. C. Chow, *J. Ceram. Soc. Jpn.*, 1991, **99**, 954.
81. T. Sugama and M. Allan, *J. Am. Ceram. Soc.*, 1992, **75**, 2076.
82. A. A. Mirtchi, J. Lemaitre and E. Munting, *Biomaterials.*, 1991, **12**, 505.
83. M. Otsuka, Y. Matsuda, Y. Suwa, J. L. Fox and W. Higuchi, *J. Biomed. Mater. Res.*, 1995, **29**, 25.
84. Y. Miyamoto, K. Ishikawa, M. Takechi, T. Toh, T. Yuasa, M. Nagayama and K. Suzuki, *Biomaterials*, 1998, **19**, 707.
85. R. P. del Real, J. C. C. Wolke, M. Vallet-Regí and J. A. Jansen, *Biomaterials*, 2002, **23**, 3673.
86. M. Nilsson, E. Fernandez, S. Sarda, L. Lidgren and J. A. Planell, *J. Biomed. Mater. Res.*, 2002, **61**, 600.
87. B. R. Constanz, I. C. Ison, M. T. Fulmer, R. D. Fulmer, R. D. Poser, S. T. Smith, M. Vanwagoner, J. Ross, S. A. Goldstein, J. B. Jupiter and D. I. Rosental, *Science.*, 1995, **267**, 1796.
88. S. Takagi, L. C. Chow and K. Ishikawa, *Biomaterials*, 1998, **9**, 1593.
89. W. S. Pietrzak and R. Ronk, *J. Craniofac. Surg.*, 2001, **11**, 327.
90. C. E. Rawlings III, R. H. Wilkins, J. S. Hanker, N. G. Georgiade and J. M. Harrelson, *J. Neurosurg.*, 1988, **69**, 269.
91. S. Sato, T. Koshino and T. Saito, *Biomaterials*, 1998, **19**, 1895.
92. M. V. Cabañas, L. M. Rodríguez-Lorenzo and M. Vallet-Regí, *Chem. Mater.*, 2002, **14**, 3550.
93. D. Yu, J. Wong, Y. Matsuda, J. L. Fox, W. I. Higuchi and M. Otsuka, *J. Pharm. Sci.*, 1992, **81**, 529.

94. C. Hamanishi, K. Kitamoto, S. Tanaka, M. Osuka, Y. Doi and T. Kitahashi, *J. Biomed. Mater. Res. Appl. Biomater.*, 1996, **33**, 139.
95. B. Mousset, M. A. Benoit, C. Delloye and R. Bouillet, *Guillard. Int. Orthop.*, 1997, **21**, 403.
96. L. Meseguer-Olmo, M. J. Ros-Nicolás, M. Clavel-Sainz, V. Vicente-Ortega, M. Alcaraz-Baños, A. Lax-Pérez, D. Arcos, C. V. Ragel and M. Vallet-Regí, *J. Biomed. Mater. Res.*, 2002, **61**, 458.
97. A. Ratier, I. R. Gibson, S. M. Best, M. Freche, J. L. Lacout and F. Rodríguez, *Biomaterials.*, 2001, **22**, 897.
98. J. C. Doadrio, D. Arcos, M. V. Cabañas and M. Vallet-Regí, *Biomaterials.*, 2004, **25**, 2629.
99. G. Daculsi, *Biomaterials*, 1998, **19**, 1473.
100. G. Drimandi, P. Weiss, F. Millot and G. Daculsi, *J. Biomed. Mater. Res.*, 1998, **39**, 660.
101. C. V. Ragel, M. Vallet-Regí and L. M. Rodriguez-Lorenzo, *Biomaterials*, 2002, **23**, 1865.
102. A. Rámila, S. Padilla, B. Muñoz and M. Vallet-Regí, *Chem. Mater.*, 2002, **14**, 2439.
103. D. C. Tancred, B. A. O. McCormack and A. J. Carr, *Biomaterials*, 1998, **19**, 2303.
104. J. M. Bouler, M. Trecant, J. Delecrin, J. Royer, N. Passuti and G. Gaculci, *J. Biomed. Mater. Res.*, 1996, **32**, 603.
105. A. Slosarczyk and J. Piekarcyk, *Ceram. Int.*, 1999, **25**, 561.
106. N. Kivrak and Tas. A. Cuneyt Tas, *J. Am. Ceram. Soc.*, 1998, **82**, 2245.
107. O. E. Petrov, E. Dyulgerova, L. Petrov and R. Ropova, *Mater. Lett.*, 2001, **48**, 162.
108. X. Yang and Z. Wang, *J. Mater. Chem.*, 1998, **8**, 2233.
109. F. H. Lin, C. J. Liao, K. S. Chen, J. S. Sun and C. Y. Lin, *J. Biomed. Mater. Res.*, 2000, **51**, 157.
110. K. Itatani, T. Nishioka, S. Seike, F. S. Howell, A. Kishiota and M. Kinoshita, *J. Am. Ceram. Soc.*, 1994, **77**, 801.
111. I. Manjubala and M. Sivakimar, *Mater. Chem. Phys.*, 2001, **71**, 272.
112. Tas. A. Cunneyt, *J. Eur. Ceram. Soc.*, 2000, **20**, 2389.
113. O. Gauthier, J. M. Bouler, E. Aguado, R. Z. LeGeros, P. Pilet and G. Daculsi, *J. Mater. Sci.: Mater. Med.*, 1999, **10**, 199.
114. E. I. Suvurova and P. A. Buffat, *Eur. Cells Mater.*, 2001, **1**, 27.
115. Y. Doi, T. Koda, N. Wakamatsu, T. Goto, H. Kamemizu, Y. Moriwaki, M. Adachi and Y. Suwa, *J. Dent. Res.*, 1993, **72**, 1279.
116. J. C. Elliot, G. Bond and J. C. Tombe, *J. Appl. Crystallogr.*, 1980, **13**, 618.
117. D. Tadic and M. Epple, *Biomaterials*, 2004, **25**, 987.
118. M. Okazaki, T. Matsumoto, M. Taira, J. Takakashi and R. Z. LeGeros, *Bioceramics 11*, R. Z. Legeros and J. P. LeGeros ed., World Scientific, New York,. 1998, 85.
119. Y. Doi, T. Shibutani, Y. Moriwaki, T. Kajimoto and Y. Iwayama, *J. Biomed. Mater. Res.*, 1998, **39**, 603.

120. M. Vallet-Regí, A. Rámila, S. Padilla and B. Muñoz, *J. Biomed. Mater. Res.*, 2003, **66**, 580.
121. R. Z. LeGeros, *Nature*, 1965, **206**, 403.
122. L. J. J. Jha, S. M. J. Best, J. C. Knowles, I. Rehman, I. D. Santos and W. Bonfield, *J. Mater. Sci. Mater. Méd.*, 1997, **8**, 185.
123. L. L. Hench, J. Wilson, L. L. Hench and J. Wilson, *An Introduction to Bioceramics.*, World Scientific, Boca Raton, FL, 1992, 20.
124. K. Ohura, T. Nakamura, T. Yamamuro, T. Kokubo, Y. Ebisawa, Y. Kotoura and M. Oka, *J. Biomed. Mater. Res.*, 1991, **25**, 357.
125. E. M. Carlisle, *Science*, 1970, **167**, 179.
126. E. M. Carlisle, *D. Calcif. Tissue Int.*, 1981, **33**, 27.
127. A. J. Ruys, *J. Aust. Ceram. Soc.*, 1993, **29**, 71.
128. Y. Tanizawa and T. Suzuki, *J. Chem. Soc. Faraday Trans.*, 1995, **91**, 3499.
129. L. Boyer, J. Carpena and J. L. Lacout, *Solid State Ion.*, 1997, **95**, 121.
130. I. R. Gibson, S. M. Best and W. Bonfield, *J. Biomed. Mater. Res.*, 1999, **44**, 422.
131. T. Kokubo, H. Kushitani, S. Sakka, T. Kitsugi and T. Yamamuro, *J. Biomed. Mater. Res.*, 1990, **24**, 721.
132. P. A. A. P. Marques, M. C. F. Magalhaes, R. N. Correia and M. Vallet-Regí, *Key Eng. Mater.*, 2001, **192–195**, 247.
133. F. Balas, J. Pérez-Pariente and M. Vallet-Regí, *J. Biomed. Mater. Res.*, 2003, **66A**, 364.
134. D. Arcos, J. Rodriguez-Carvajal and M. Vallet-Regí, *Phys. Rev. B.*, 2004, **350**, e607.
135. M. Vallet-Regí, *Anales de Quím. Inter.* l Ed. Suplement 1. 1997, **93.1**, S6.
136. S. H. Shn, H. K. Jun, C. S. Kim, K. N. Kim, S. M. Chung, S. W. Shin, J. J. Ryu and M. K. Kim, *J. Oral Rehab.*, 2006, **33**, 12.
137. L. Tuck, M. Sayer, M. Mackenzie, J. Hadermann, D. Dunfield, A. Pietak, J. W. Reid and A. D. Stratilatov, *J. Mater. Sci.*, 2006, **41**, 4273.
138. E. S. Thian, J. Huang, S. M. Best, Z. H. Barber and W. Bonfield, *J. Biomed. Mater. Res.*, 2006, **78A**, 121.
139. J. Peña, R. P. del Real, L. M. Rodríguez-Lorenzo and M. Vallet-Regí, In *Bioceramics. 12.*, H. Ohgushi, G. W. Gastings and T. Yoshihawa ed., World Scientific Publishing Co. Pte. Ltd, Nara, Japan, 1999, 353.

CHAPTER 3

Biomimetic Nanoapatites on Bioceramics

3.1 Introduction

Biomimetic materials science is an evolving field that studies how Nature designs, processes and assembles/disassembles molecular building blocks to fabricate high-performance minerals, polymers and mineral-polymer composites
(*e.g.*, mollusc shells, bone, tooth) and/or soft materials (*e.g.*, skin, cartilage,
tendons) and then applies these designs and processes to engineer new molecules and materials with unique properties.[1] The fabrication of nanostructured
materials that resemble the complex hierarchical structures of natural hard
tissues present in bones and teeth is a primary objective from the point of view
of *biomaterials science*. We have seen in Chapter 1 that bone is an excellent
example of hierarchical organisation with structural and functional purposes,
where the transition from the nanometric to the macroscopic scale is carefully
organised.[2] However, the development in biomaterials science is still far away
from this objective, and perhaps a more realistic aim is to design implant
surfaces at the nanometric scale to optimise the tissue/implant interface,[3]
facilitating the bone self-healing.

   In Chapter 2, we could see how bone-like HA (hydroxyapatite) nanoparticles
can be synthesised by a range of production methods, such as precipitation
from aqueous solutions, sol-gel synthesis, aerosol assisted methods, *etc*. In this
chapter we will deal with one of the most promising and developed methods:
the *biomimetic synthesis*. In the frame of the *bioceramics field*, biomimetism is
considered as mimicking natural manufacturing methods to generate artificial
bone like calcium phosphates, mainly apatites, which can be used for bone- and
teeth-repairing purposes. The most common process consists of the crystallisation of nonstoichiometric carbonate hydroxyapatite (CHA) from simulated
physiological solutions at temperatures similar to those in physiological

RSC Nanoscience & Nanotechnology
Biomimetic Nanoceramics in Clinical Use: From Materials to Applications
By María Vallet-Regí and Daniel Arcos
© María Vallet-Regí and Daniel Arcos, 2008

conditions.[4] The bone-like apatite crystallisation takes place through the nucleation of calcium phosphate (CaP) precursors, such as amorphous calcium phosphate (ACP) or octacalcium phosphate (OCP).[5,6] These precursors subsequently mature to calcium-deficient hydroxyapatite (CDHA) by incorporating CO_3^{2-}, OH^-, Ca^{2+}, PO_4^{3-}, *etc.* ions from the surrounding solution.

The idea of using *bioactive ceramics* as substrates for biomimetic synthesis of nanoapatites acquired great importance, when in 1971 Hench *et al.*[7] discovered that the bioactive process in SiO_2-based bioceramics took place through the formation of a carbonate-containing CDHA at the implant tissue surface. Thereafter, it could be seen that the prior *in vitro* biomimetic growth of a nanocrystalline CDHA allowed the fabrication of implants with fitted-out surfaces to be colonised by bone cells.[8,9] Bone cells have been shown to proliferate and differentiate on these apatite layers, showing increased bioresponse and new bone formation.[10,11] Nowadays, among the different concepts for fabrication of highly bioresponsive nanoceramics, biomimetic methods are one of the most developed strategies to produce body interactive materials, helping the body to heal and promoting tissue regeneration. In this sense, bioceramics such as bioactive glasses, glass-ceramics and calcium-phosphate-based synthetic compounds are excellent substrates that develop calcium phosphate nanoceramics with almost identical characteristics to the biological ones, when soaked in solutions mimicking physiological conditions.

3.1.1 Biomimetic Nanoapatites and Bioactive Ceramics

The motivation to carry out the synthesis of nanostructured apatites over bioceramic surfaces arises from the understanding of the physical-chemical and biological processes that lead to the bond formation between bones and implants. When bioactive ceramics such as bioglass, apatite-wollastonite glass ceramic or HA/β–TCP biphasic calcium phosphate are implanted in bone tissue, the examination of the implant site reveals the presence of a nanocrystalline calcium-deficient carbonate apatite at the bonding interface.[12] This intermediate apatite layer is similar to biological apatites in terms of calcium deficiency and carbonate substitutions, and it was believed that it would interact with osteoblast in a similar manner as biological apatites do. On the other side, when the so-called bioactive ceramics are soaked in *artificial* or *simulated physiological fluids*, the surface analysis evidence the setting off of chemical reactions at the material surface, such as dissolution, precipitation, ionic exchange, *etc.* together with biological material adsorption.

One of the most important works evidencing the role of the newly formed apatite layer on bioactive ceramics, was carried out by Prof. Kokubo's research team.[13,14] Kokubo and coworkers systematically demonstrated that the *in vivo* bioactivity of a material, as measured by the rate of bone ingrowth, could be directly related to the rate at which the material forms apatite *in vitro* when immersed in simulated body fluids (SBF). This work consisted of synthesising different compositions of glass particles in the system Na_2O-CaO-SiO_2 that

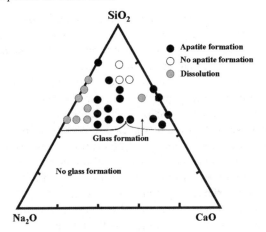

Figure 3.1 Compositional dependences of nanoapatite formation on glasses in the system Na_2O-CaO-SiO_2, after soaking in simulated body fluid.

were packed into the bony defects of rabbit femoral condyle to evaluate their ability to induce bone in-growth, while the same set of bioglass formulations (see Figure 3.1) were also immersed into simulated body fluids to evaluate the apatite-forming ability *in vitro*.

The results showed that the *in vivo* bioactivity was precisely reproduced by the apatite-forming ability *in vitro*. The glass formulation that induced apatite formation most efficiently and rapidly *in vitro* also stimulated the most significant bone-formation activity 3 and 6 weeks after implantation *in vivo*. Nowadays, the *in vitro* nucleation and growth of a nanocrystalline apatite onto a bioceramic is considered as a clear sign of a good *in vivo* behaviour. The implant–bone bonding ensures the materials osteointegration and, very often also promotes the bone-tissue regeneration. In the last cases, gene activation, implant resorption and bone-ingrowth mechanisms are also involved.

3.1.2 Biomimetic Nanoapatites on Nonceramic Biomaterials. Two Examples: Polyactive® and Titanium Alloys

The formation of nanocrystalline apatites at the implant surface sets off the bioactive bonding and/or bone-tissue regeneration when implants are in contact with living tissues. Clear examples of the biomimetic apatite layer significance can be found not only in the case of ceramic compounds, but also in polymers and metals such as Polyactive® and titanium alloys. Polyactive® is a member of a series of segmented copolymers based on polyethylene oxide and polybutylene terephthalate.[15] This polymer is considered as a potential bone substitute material with bioactive properties and, consequently, with bone-bonding ability.[16,17] The capability of Polyactive® as a potential bone substitute had been investigated with different animal models,[18–22] but some studies raised a concern about the clinical usage of this polymer, concerning its osteoconductive properties. The problem was tackled by carrying out a biomimetic growth of bone-like apatite coating, in order to stimulate or enhance the bone bonding with this polymer.[23] After being implanted in rabbit femur, abundant new bone growth with spongy appearance along the implant surface was observed after 2 weeks, and the marginal bone formation with a maximal penetration depth of about 1 mm in 4-mm diameter defects was observed after 8 weeks.

The biomimetic growth of nanoapatites has been also extended to metal alloys commonly used in orthopaedic surgery, for instance titanium alloys. The application of titanium alloys in artificial joint replacement prostheses is mainly motivated for their lower modulus, superior biocompatibility and enhanced corrosion resistance when compared to more conventional stainless steels and cobalt-based alloys. When this material is manufactured as HA-coated joint implants, the plasma spraying technique is the process commonly used in their production.[24,25] However, this technique exhibits several drawbacks related with the coating thickness heterogeneity, weak adherence and structural integrity as well as coating delamination, which lead to the fibrous tissue ingrowth and occasional implant loosening. Li[26]-implanted multichanneled Ti6Al4V implants in which four channels were apatite coated in an aqueous solution formulated to include HCO_3^- ions and other major inorganic ions present in the body such as HPO_4^{2-}, Ca^{2+}, Mg^{2+}, Na^+, K^+, Cl^- and SO_4^{2-}, which could induce the formation of an apatite coating closely mimicking bone mineral, whereas the rest of the channels remained noncoated. Eight weeks after implantation into the distal femur of dogs, the histological examination revealed much higher bone in-growth through the apatite-lined channel of all implants, while the noncoated channel had minimal in-growth.

3.1.3 Significance of Biomimetic Nanoapatite Growth on Bioceramic Implants

The improved clinical performance of Polyactive® and Ti alloys when coated with biomimetic nanoapatite, gives away the potential of this field in orthopaedic and dental surgery. Both substrates are suitable to be coated after different chemical treatments aimed to prepare their surfaces for nanoapatite crystallisation. However, in the case of bioactive ceramics, these materials not only can be coated by a newly formed CaP layer, but they strongly promote the biomimetic process and their potential for bone-tissue regeneration deserves a special attention.

The pioneering studies of Hench *et al.*[7] on the bioactive processes in SiO_2-based bioceramics, and the correlation established by Kokubo *et al.*[4] between the biomimetic nanoapatite formation and their *in vivo* performance, led to the extended use of this procedure to measure the level of bioceramics bioactivity, by examining the *in vitro* apatite-forming ability on its surface. However, after considering the advantages of the advanced nanoceramics, these nanoscaled coatings are being used to produce bioceramics with better hard- and soft-tissue attachment, higher biocompatibility and enhanced bioactivity for bone-regenerative purposes. The biological mechanisms that rule these enhanced characteristics are not fully defined. However, several performance guidelines of the biomimetic nanoapatites can be addressed[27] (Figure 3.2):

a) *In vivo* dissolution of the biomimetic nanoapatite, leading to the saturation of surrounding fluids and thus accelerating the precipitation of truly biological apatites onto the coated implant.
b) Adsorption of large amounts of protein from the neighbouring environment due to the surface charge of the nanoapatite, thus triggering cell differentiation.
c) The microstructure of the substrate/apatite coating increases the surface roughness, which is beneficial for osteoinduction as compared to smooth surfaces.

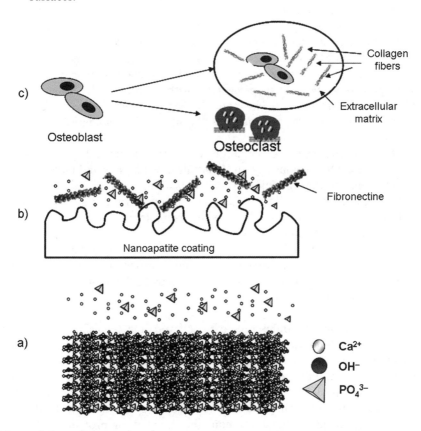

Figure 3.2 Possible biological mechanisms that govern the improved response of coated bioceramic surfaces with biomimetic nanoapatites. a) Dissolution of the biomimetic nanoapatite leading to the saturation of surrounding fluids. b) Increase of the surface roughness and adsorption of large amounts of cell adhesion proteins. c) Ca^{2+} and PO_4^{3-} ions may signal cells toward the osteoblast differentiation.

    d) The apatite could be the source for Ca^{2+} and PO_4^{3-} ions that may signal cells toward the differentiation pathway and trigger bone formation.

    e) Since the biomimetic nanoapatite is similar in structure and properties to natural biological apatites, it could constitute an excellent substrate for new biological phase nucleation.

3.2 Simulated Physiological Solutions for Biomimetic Procedures

Nanocrystalline hydroxyapatite coatings can be easily produced on various ceramic substrates through the reaction with artificial physiological fluids. In general terms the biomimetic formation of apatite involves nucleation and growth from an ionic solution.[28,29] The composition of any solid deposited on the surface of a bioceramic will be largely determined by the surrounding media, so choosing the correct experimental conditions and mimicking solution is mandatory. The apatite crystallisation from a solution could be reached by mixing aqueous solutions containing the calcium and phosphates ions. However, this kind of process would lead to precipitates with properties very different with respect to biological apatites. There are obvious differences between the *in vivo* and the *in vitro* crystallisation conditions,[30] which can be summarised as follow:

1. Depleting concentration conditions commonly occur under *in vitro* crystallisation. On the contrary, the concentrations of ions and molecules are kept constant during biological mineralisation.
2. Kinetics of the precipitation reaction. Chemical crystallisation is a much faster process (minutes to days), while the biological process is measured in terms of weeks and even years.
3. Presence of inorganic, organic, biological and polymeric compounds within biological fluids, which are commonly absent in artificial solutions. These species often act as inhibitors, seeds and templates during the growth of biological apatites.

The first and second differences can be overcome by using appropriate crystallisation techniques and this topic will be discussed later. However, the third requires a more complex approximation, involving chemical and biological concepts, to fabricate appropriate crystallisation solutions. Using natural fluids such as blood, saliva, *etc.* involves serious drawbacks related to the amounts available, variability and storage. Moreover, in the case of solutions able to mimic bone apatite formation, the presence of proteins and other biological entities exert a high inhibitory or delaying effect.[31-33] For this reason, inorganic ionic solutions are the most widely applied fluids for biomimetic nanoapatite purposes.

Among the different artificial solutions able to partially simulate the physiological conditions, the simulated body fluid (SBF) developed by Prof. Kokubo is

Table 3.1 Human plasma and ion concentration of some of the most applied artificial solutions for biomimetic processes (mM).

	Na^+	K^+	Ca^{2+}	Mg^{2+}	HCO_3^-	Cl^-	HPO_4^{2-}	SO_4^{2-}
Human plasma (total)	142.0	5.0	2.5	1.5	27.0	103.0	1.0	0.5
Human plasma (dissociated)	142.0	5.0	1.3	1.0	27.0	103.0	1.0	0.5
SBF	142.0	5.0	2.5	1.5	4.2	148.0	1.0	0.5
i-SBF	142.0	5.0	1.6	1.0	27.0	103.0	1.0	0.5
m-SBF	142.0	5.0	2.5	1.5	10.0	103.0	1.0	0.5
r-SBF	142.0	5.0	2.5	1.5	27.0	103.0	1.0	0.5
n-SBF	142.0	5.0	2.5	1.5	4.2	103.0	1.0	0.5
HBSS[45]	142.0	5.8	1.3	0.8	4.2	145.0	0.8	0.8
PECF[46]	145.0	5.0	–	–	30.0	118.0	1.0	–
EBSS[47]	144.0	5.4	1.8	0.8	30.0	125.0	1.0	–
PBS[48,49]	146.0	4.2	–	–	–	141.0	9.5	–

the most widely applied solution for biomimetic purposes.[4] SBF is a metastable aqueous solution with pH of around 7.4, supersaturated with respect to the solubility product of HA. This solution only contains inorganic ions in concentration almost equal to the human plasma (Table 3.1). The main difference between SBF and the inorganic part of the biological plasma is the bicarbonate (HCO_3^-) concentration, which is significantly lower in SBF (4.2 mM instead of 27 mM in plasma). SBF has been widely used for *in vitro* bioactivity assessment of artificial materials by examining their apatite-forming ability in the fluid.[34–36] On the other hand, SBF has also been used to prepare artificial bone-like apatite on various types of substrates.[37–39] In this sense, controlling the composition and structure of the apatite produced in SBF has been one of the most important aims in the framework of biomimetic synthesis, and several efforts have been made in order to precipitate apatites equal (or very similar) to those occurred in bones.

Kim *et al.*[40,41] reported that the apatite produced in a conventional SBF differs from bone apatite in its composition and structure. They attributed this difference to the higher Cl^- and lower HCO_3^- concentrations of the SBF than those of blood plasma, (see Table 3.1), and they demonstrated that an apatite with a composition and structure similar to that of bone would be produced if the SBF had ion concentrations almost equal to those of human plasma. When tailoring new SBFs, it must be taken into account that of the calcium ions in blood plasma (2.5 mM), 0.9 mM of Ca^{2+} are bound to proteins, and 0.3 mM of Ca^{2+} are bound to inorganic ions, such as carbonate and phosphate ions.[42] Considering this, Oyane *et al.*[43] prepared new SBFs denoted:

- Ionised SBF (i-SBF), designed to have concentrations of dissociated ions equal to those of blood plasma.
- Modified SBF (m-SBF), designed to have concentrations of ions equal to those of blood plasma, excepting HCO_3^-, the concentration of which is decreased to the level of saturation with respect to calcite ($CaCO_3$).
- Revised SBF (r-SBF), designed to have a concentration of ions all of which are equal to those of blood plasma, including Cl^- and HCO_3^-.

The main drawback of i-SBF and r-SBF is their stability. These two fluids are less stable than SBF and m-SBF in terms of calcium carbonate cluster formation. For these reasons, these two fluids are not suitable for long-term use in bioactivity assessment of materials, although they can be used for biomimetic synthesis of bone-like apatite. On the other hand, m-SBF is stable for a long time with respect to changes in ion concentrations and, in contact with bioceramics, the m-SBF better mimics the biological apatite formation compared with conventional SBF.

In 2004, Takadama *et al.*[44] proposed a newly improved SBF (n-SBF) in which they decreased only the Cl^- ion concentration to the level of human blood plasma, leaving the HCO_3^- ion concentration equal to that of the conventional SBF (SBF). n-SBF was compared with conventional SBF in terms of stability and reproducibility of apatite formation, evidencing that SBF does not differ from n-SBF and both solutions could be indifferently used for biomimetic studies.

Further attempts to improve the biomimetic properties of SBF have been performed. Some efforts have been made to replace artificial buffers by simultaneously increasing the hydrogen carbonates concentration of SBF or avoiding CO_2 losses from SBF through the permanent bubbling of CO_2. Addition of the most important organic and biological compounds such as glucose and albumin is another direction to improve biomimetic properties of SBFs, although the presence of proteins can seriously impede the HA crystallisation.

Occasionally, condensed solutions of SBF ($\times 1.5$, $\times 2$, $\times 5$ and even $\times 10$ concentration) are used to accelerate the precipitation;[41,50–54] the use of condensed solutions is controversial since it leads to changes in the chemical composition of the biomimetically growth calcium phosphate. Commonly, the crystallised apatite exhibits different microstructures and lower phosphate amounts due to a higher carbonate ions incorporation, which could affect to the osteoblast response when a biomimetic nanoapatite makes contact with them. This effect has been studied onto culture-grade polystyrene.[55] For instance, biomimetic treatments of 1 day into SBF followed by 14 days in more concentrated SBF (SBF $\times 1.5$) lead to nanocrystalline Ca deficient carbonatehydroxyapatite (CHA), *i.e.* conventional biomimetic apatite commonly observed on the surface of bioactive ceramics after a few days in SBF. Nanocrystalline octacalcium phosphate (OCP) or even amorphous calcium phosphate (ACP), considered as HA precursors during the biomineralisation process, can be obtained at the implant surface by homogeneous precipitation within highly supersaturated SBF (SBF $\times 5$). This kind of solution cannot be prepared at a physiological pH of 7.4, and acid pH values are required to avoid immediate precipitation. Once these precursors are formed, they can be converted into biomimetic apatite by soaking the substrates in SBF depleted of HA crystal growth inhibitors, *i.e.* without Mg^{2+} and HCO_3^-. At this point, it is important to highlight that, depending on the pH at which the precursors were precipitated, the microstructure of the final HA can vary from large plate-shaped crystals (CaP precursor precipitated at pH around 6.5) to small platelet

Table 3.2 Different treatments for synthesising biomimetic apatites and their effect on the osteoblastic response.

Simulated fluid treatment	Biomimetic CaP precipitated	Osteblastic cell response
SBF (1 day) + 1.5× SBF (14 days)	Conventional biomimetic CHA	Good development of anchoring elements Osteoblast elongation
5–SBF (pH 5.8) *or* 5–SBF (pH 6.5), 1 day	HA precursors (OCP or ACP)	High cellular death
5–SBF (pH 5.8), 1 day + SBF depleted of Mg^{2+} and HCO_3^-, 2 days	Small plate CHA	Cell viability, narrowing of anchoring elements, better spreading degree
5–SBF (pH 6.5), 1 day + SBF depleted of Mg^{2+} and HCO_3^-, 2 days	Large plate CHA	Enhanced formation of extracellular matrix and biomineralisation process. Enhanced cell differentiation

crystallites (CaP precursor precipitated at pH below 6). Table 3.2 summarises the conditions to control the biomimetic process through the combination of solutions.

Table 3.2 shows how the chemical composition and microstructure of the biomimetic CaP is determining for an appropriated osteoblastic cells response. Viability *in vitro* cell culture studies indicate that biomimetic CaP precursors lead to higher percentages of cellular death, especially during the first days of culture. This fact seems to be related with the high reactivity of the biomimetically formed OCP or ACP precursors with the culture media, leading to strong microenvironmental changes in the calcium and phosphate ions concentrations. On the contrary, biomimetic HA enhances the formation of extracellular matrix (ECM) and the biomineralisation process by the osteoblastic cells. Moreover, when osteoblasts are seeded onto biomimetic HA there is an enhanced expression of osteocalcin and bone sialoprotein – ECM mineralisation markers – compared with polystyrene substrates, especially in those media depleted of inductive agents of osteoblastic gene expression such as exogenous ascorbic acid and β-glycerol phosphate. The phosphorus presence at microenvironment of biomimetic surfaces seems to provoke this response.

Finally, biomimetic HA enhances the cell differentiation as deduced from the higher expression of osteopontin mRNA, especially large HA platelets. The mechanism is not still clear, but the better protein adsorption points out that integrin-mediated signalling would be involved in the process.

Although SBF is a very useful fluid to mimic the "inorganic events" that occurred during the bioactive process *in vivo*, the high ionic saturation makes the study of dissolution, precipitation and ionic exchange processes between the fluid and the ceramic difficult. For this reason simpler solutions such as tris(hydroxymethyl) aminomethane buffered solution at pH 7.3 are often preferred to determine the bioactive behaviour of bioceramics like bioglass[56] especially for those studies where ion kinetic dissolution is the main focus of the research.

Table 3.1 also displays other solutions commonly used for biomimetic purposes. Hanks and Wallace[57] balanced salt solution (HBSS) was the first successful simulated medium, containing the ions of calcium and phosphates together with other inorganic ions and glucose. HBSS is commercially available and still used in biomimetic experiments.[58,59] Homsy's pseudoextracellular fluid (PECF) is another phosphate containing solution that also used for biomimetic apatite growth. Earl's balanced salt solution (EBSS) is a tissue culture medium that contains varying amounts of $CaCl_2$, $MgSO_4$, KCl, $NaHCO_3$, NaCl, $NaH_2PO_4.H_2O$ and glucose, according to the application and technique. It is commercially available in premixed salts or in solution. Finally, phosphate-buffered saline (PBS) is a buffer solution commonly used in biochemistry. It is also a commercially available solution that only contains inorganic components and is suitable for biomimetic purposes.

3.3   Biomimetic Crystallisation Methods

In principle, the biomimetic coating procedures and the bioactivity tests described above involve a solution that is not renewed. Thus, the ions released from the glass remain in the container. This method is termed *static* or *integral*[60] and it is widely accepted that monitoring the formation of a CHA layer in these conditions predicts the material's bioactive behaviour. However, the use of the *static procedure* with highly reactive materials in aqueous solutions leads to remarkable variations in the ionic concentration and pH, reaching values far from physiological ones. This fact makes questionable the accuracy of these assays or the similarity of the coating with respect to biological apatites. Increases of pH of around 0.6 units from the initial 7.4 can be observed in the SBF, when bioactive sol-gel glasses are soaked for a few hours. Besides, variations in the ionic concentration of Ca(II), P(V) and Si(IV) are also detected just after a few minutes of assay.[61] Such pH increases could favour the CHA formation even in weakly bioactive materials.

Some authors have proposed the so-called *differential* method[62,63] in which the solution is renewed at predetermined intervals. However, the periodical solution exchange to eliminate such effects in bioactive glasses would require such short time intervals that the formation process of the CHA layer could be affected by the sample manipulation.

For that reason, also to simulate the continuous flux of body fluids at the implant surface, *dynamic* or *continuous in vitro* procedures have been proposed,[64] in which SBF is continuously renewed with the aid of a peristaltic pump. Figure 3.3 shows the scheme of the device used for *dynamic in vitro* assays.

Dynamic tests have been used to assess the *in vitro* bioactivity of several glasses, and compared with that without the renewal of the *in vitro* solution (*static*). A SBF flux at 1 mL/min allows the ionic concentration and pH of the solution to be maintained almost constant. As expected, the protocol modifications result in variation of the nanoapatite growth from both chemical and

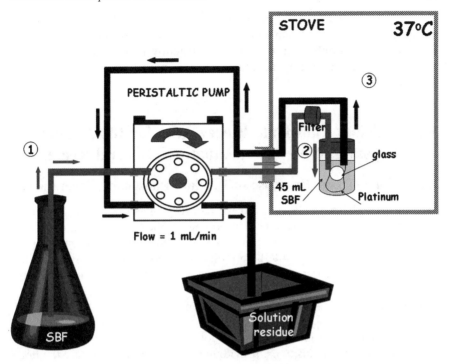

Figure 3.3 Schematic description of the dynamic *in vitro* bioactivity assays. The continuous flow of the body fluids is modelled by the continuous renewal of the SBF solution.

microstructural point of view. In *static* conditions, a faster initial formation of the amorphous phosphate coating is detected, but for higher soaking times the situation is equivalent in both cases. Under *dynamic* conditions, the formed apatite crystals are larger. Regarding the layer composition in *dynamic* conditions, the Ca/P molar ratio is considerably lower than in the *static* case (1.2 *vs.* 1.6). This variation was explained by the differences in pH. The lower pH in *dynamic* conditions (7.4) increases the HPO_4^{2-} concentration in solution compared with *static* where pH is close to 8. Thus, *dynamic* would favour the formation of calcium-deficient apatite, which might coexist with other calcium phosphates of lower Ca/P molar ratio. In addition, the larger size of the CHA crystals aggregates formed under *dynamic* conditions is explained on the basis of the continuous supply of calcium and phosphate ions.

Other alternatives are the use of constant composition techniques such as those proposed by Nancollas *et al.*[65–67] In these methods, multiple titrant solutions containing lattice ions are added to the reaction solution to compensate for the removal of these ions during growth. Thus, a constant thermodynamic driving force for crystal growth is maintained during the calcium phosphate growth. In order to mimic the kinetics of biological apatite crystallisation, other methods such as a double-diffusion crystallisation device or crystallisation within viscous gels have been proposed.[68–72] These methods are based on the restrained diffusion of calcium and phosphate ions from the opposite direction. Together with a double-diffusion process currently they are considered one of the most advanced experimental tools for mimicking biomineralisation processes.

3.4 Calcium Phosphate Bioceramics for Biomimetic Crystallisation of Nanoapatites. General Remarks

3.4.1 Bone-Tissue Response to Calcium Phosphate Bioceramics

Calcium phosphates (CaP) fall into the category of biocompatible materials for bone and dental applications. Depending on their chemical composition, crystalline phase and microstructure, CaPs can slightly dissolve, promoting the formation of biological apatite before directly bonding with the tissue at the atomic level. This process results in the formation of a direct chemical bond with bone and it is named *bioactivity*.

After implantation, CaPs can act in different ways:

1) *Osteoconduction*. Giving rise to a good stabilisation through an osteo-conductive mechanism, *i.e.* providing a bioactive surface where the bone can grow on without implant resorption.
2) *Osteoinduction*. Osteoinductive materials will stimulate the osteoblast proliferation and differentiation by providing biochemical signals that result in bone-tissue regeneration. Osteoinduction is a property not tra-ditionally attributed to calcium phosphate ceramics, but recent studies have demonstrated osteoblast stimulation for several CaP compositions.[73]
3) *Bioresorption*. A bioresorbable material will dissolve and allow a newly formed tissue to grow into any surface irregularities but may not neces-sarily interface directly with the material. In the field of calcium-phos-phate-based bioceramics, we can find examples of all the situations described above.[74,75]

Independently of the chemical composition, structure and microstructure of a bioactive ceramic, the analysis of the bone/implant interface reveals that the *presence of nanocrystalline calcium-deficient hydroxyapatite (CDHA)* is one of the key features in the bonding zone.[7,76] In the case of CaP bioceramics, a second *rule* can also be established: *the implant solubility enhances the bone-repair process*.[77–80] It does not mean that only highly soluble CaP is useful for bone repairing; CaPs with higher solubility are applied in those applications where the implant resorption is expected, followed by bone colonisation, whereas less-soluble CaPs are intended as osteoconductive materials, providing a bioactive surface that supports bone growth without dissolving, with better mechanical stability during the first stages of the repairing process. Bioceramics made of dense HA would be a good example of bioactive material,[81,82] while porous scaffolds made of biphasic calcium phosphate, BCP, β-TCP/HA[83] or α–TCP/HA[84]) or bone grafts made of CDHA or ACP are examples of bio-resorbable materials.[85,86]

3.4.2 Calcium Phosphate Bioceramics and Biological Environment. Interfacial Events

The ability to bond to bone tissue is a unique property of bioactive materials. During this process, dissolution and precipitation reactions occur. Figure 3.4 schematically shows these phenomena, with a list of events occurring during the bioactive process. The events that constitute the bioactive process are commonly overlapped or simultaneously occurring, and the scheme displayed in Figure 3.4 should not be considered in terms of a time sequence.[87]

The scheme displayed in Figure 3.4 does not represent a mechanism by itself, but only a description of observable events that occur at the interface after implantation. The mechanism must be related with physicochemical phenomena that occur in the presence or absence of cells, or are related to reactions mediated by cellular activity. An important aspect of the overall reaction sequence between these materials and tissues is that, in the absence of biologically equivalent calcium-deficient carbonate apatite, dissolution, precipitation and ion-exchange reactions lead to a biologically equivalent apatitic surface on the implanted material: *the in vivo bioactivity is only strongly expressed if this new calcium-deficient carbonate apatite is formed.* Under *in vitro* conditions in noncellular simulated physiological conditions, the stages 1 to 4 are reproduced, leading to the precipitation of biomimetic calcium phosphates.

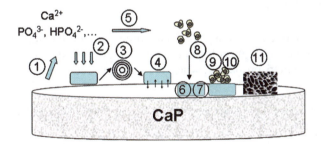

Figure 3.4 Events occurring during bone formation onto bioactive CaP ceramics. 1) Dissolution from the ceramic. 2) Precipitation from solution onto the ceramic. 3) Ion exchange and structural rearrangement at the ceramic/tissue interface. 4) Interdiffusion from the surface boundary layer into the ceramic. 5) Solution-mediated effects on cellular activity. 6) Deposition of either the mineral phase or the organic phase, without integration into the ceramic. 7) Deposition with integration into the ceramic. 8) Chemotaxis to the ceramic surface. 9) Cell attachment and proliferation. 10) Cell differentiation. 11) Extracellular matrix formation.

3.4.3   Physical-Chemical Events in CaP Bioceramics during the Biomimetic Process

3.4.3.1   CaP Dissolution during Biomimetic Processes

The reactivity of a CaP is dependent on its composition and structure.[88] One of the mechanisms underlying the phenomena of *in vitro* bioactivity is that dissolution from the ceramic produces solution-mediated events leading to mineral precipitation.[77,78] Under *in vivo* conditions, the process involves more complicated biological reaction affecting cellular activity and organic matrix deposition.[89,90]

Table 3.3 displays some chemical and textural characteristics of the most important CaP bioceramics in the field of dental and orthopaedic surgery. When studying the physical-chemical features of CaP compounds, we must take into account the following parameters:

1. Type of calcium phosphate ceramic, *i.e.* hydroxyapatite, tricalcium phosphate, tetracalcium phosphate, *etc.*
2. Type of crystal-chemical defects, such as deviation from stoichiometry leading, for instance, to calcium-deficient compounds, dehydroxylation, *etc.*
3. Polymorph considered for a chemical compound, such as α-TCP and β-TCP.
4. Number and type of CaP phases existing in the system, commonly monophasic or biphasic CaP systems.

Even considering all these parameters, the question about the CaP solubility under the action of a physiological solution is not trivial. In order to quantify the dissolution of a CaP when soaked into a buffered fluid, two different approaches can be applied: *Initial dissolution rate* and *Concentration of dissolved ions at the equilibrium.*[91]

3.4.3.1.1   Initial Dissolution Rate   Determination of the initial dissolution rate must be carried out with the data points experimentally obtained at the short initial immersion period, when the ionic product does not vary or does

Table 3.3   Crystalline phases, Ca/P ratio and surface area of some calcium phosphate bioceramics.

Bioceramic	Phases	Ca/P ratio	S_{BET} (m^2/g)
HA	Stoichimetric HA	1.67	5.1
CDHA	Ca-deficient HA	1.61	62.9
OHA	Oxyhydroxyapatite	1.67	2.48
β-TCP	β-tricalcium phosphate	1.5	0.64
α-TCP	α-tricalcium phosphate	1.5	0.08
TTCP	Tetracalcium phosphate	2.0	0.24
BCP-45	45 HA/55 β-TCP	1.58	5.05
BCP-27	27 HA/73 β-TCP	1.55	4.15

not significantly affect the undersaturation factor. Under these conditions, the initial dissolution rate is related to the following rate expression

$$\frac{d[Ca]}{dt} = k \cdot t^m \tag{3.1}$$

with

[Ca]: Ca concentration in solution
k: constant
m: effective order of the reaction

Developing the derivative function, the initial dissolution rate can be expressed in an easy logarithmic expression as a function of soaking time, as follows:

$$\log[Ca] = A_0 + A_1 \log t \tag{3.2}$$

where

$$A_0 = \log(k/m + 1)$$

$$A_1 = m + 1$$

In this way, by measuring the Ca^{2+} concentration as a function of soaking time, the solubility of the CaP substrate can be determined attending to some specific characteristic. For instance, the influence of crystal-chemical defects can be estimated by comparing the solubility of stoichiometric HA, partially dehydroxylated oxyhydroxyapatite (OHA) and CDHA. Experimentally, it can be observed that solubility increases in the order

$$HA < CDHA < OHA$$

Following the same procedures, it was observed that factors such as high specific surface area, crystallographic defects and nonstoichiometry enhance the dissolution rate of the CDHA.[92] The general formula for CDHA is: $Ca_{10-x}(HPO_4)_x(PO_4)_{6-x}(OH)_{2-x}$, where x can vary from 0 to almost 2.[93] In addition to the Ca deficiency, the low carbonate content generally contained in these compounds, contributes to the structural disorder of the CDHA by replacing the tetrahedral PO_4 group by a planar CO_3. Finally, the Ca deficiency is also accompanied by hydroxyl-group deficiency. This set of crystal-chemical defects leads to the higher initial dissolution rate of CDHA when compared to HA.

OHA can be presented by a formula: $Ca_{10}(PO_4)_6(OH)_{2-2x}O_x\square_x$, where $\square$ means a vacancy. In the case of OHA, one O^{2-} ion and a vacancy substitute for two monovalent OH^- ions. The enhanced Ca^{2+} release would be a consequence

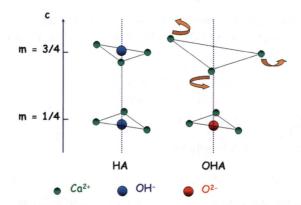

Figure 3.5   Scheme of the ions along the *c*-axis in the hydroxyapatite (left) and oxy-
                 hydroxyapatite (right). The vacancies at the hydroxyl sites in OHA result
                 in weaker ionic interactions with the Ca^{2+} facilitating the ion dissolution.

of the weak bonding interaction of Ca^{2+} ions around the vacancies, as is
schemed in Figure 3.5.

Regarding phosphate release, CDHA also shows a higher dissolution rate
compared with HA. The same factors contributing to Ca^{2+} release can explain
the phosphate dissolution. On the contrary, OHA does not show a P release
enhancement with respect to HA. Probably, the lower amount of OH^- groups
decreases the hydrogen attraction on its surface. Since H^+ governs the solid to
solution exchange of the phosphate ions, this crystal-chemical feature could be
responsible for the lower P release in the case of OHA.

Obviously, different dissolution rates are observed when compared to different CaP phases. Greater Ca^{2+} and PO_4^{3-} release rate from β-TCP than from
HA is expected, since β-TCP is known as a metastable member of the CaP
family.[94] β-TCP cannot be precipitated from aqueous solutions, but it is a high-
temperature phase of calcium orthophosphates, which only can be prepared by
thermal decomposition, *e.g.* of CDHA, at temperatures above 800 °C. As well
as exhibiting lower surface areas than HA, β-TCP shows a larger dissolution
rate than HA. At temperatures above 1125 °C, transformation of β-TCP to α-
TCP takes place. α-TCP is more soluble in aqueous media and both the initial
dissolution rate and the ionic product for α-TCP are significantly greater than
those for β-TCP. Excellent and extensive information on this topic can be
found in the books of Elliott[95] and LeGeros[96].

Among the considered single-phase CaPs, TTCP shows one of the greatest
initial dissolution rates. However, a rapid increase in Ca and P content is
commonly followed by a decrease first in P content and then in Ca content.
It indicates that the solution with immersed TTCP became rapidly saturated
with one of the metastable CaPs phases. The subsequent decrease of the P and
Ca content is the result of precipitation of new phase(s) on the TTCP surface.

Biphasic calcium phosphates (BCPs) consist of mixtures of HA and
β-tricalcium phosphate (β-TCP). Due to the higher solubility of the β-TCP

component, the reactivity increases with the β-TCP/HA ratio. Therefore, the bioreactivity of these compounds can be controlled through the phase composition.

The initial dissolution rates of the single-phase CPCs in undersaturated conditions at physiologic pH increase in the order:

$$HA < CDHA < OHA < \beta\text{-TCP} < \alpha\text{-TCP} < TTCP$$

whereas BCPs solubility would fall somewhere between HA and β-TCP, depending upon the quantitative phase composition.

3.4.3.1.2 Concentration of Dissolved Ions at the Equilibrium Since dissolved ions are transported away by the physiological fluids under *in vivo* conditions, the concentration of dissolved species at equilibrium is not a useful parameter to explain the bioactive behaviour of bioceramics. However, when considering the *in vitro* biomimetic synthesis of nanoapatites, it becomes an essential parameter to understand the subsequent nanoapatite precipitation, especially when *integral* methodology is applied.

Table 3.4 displays some of the CaP ceramics with application in dental and orthopaedic surgery, together with the solubility parameters and pH stability.

The precipitation of CaPs is known to be principally determined by calcium and phosphate concentrations and condition of the nucleation site. Therefore, the amount of bioceramic dissolved at the equilibrium point strongly depends on the ionic strength of the solution.

In the case of CaP-based bioceramics, reaching the saturation points for Ca^{2+} and phosphates is very important for the biomimetic formation of calcium phosphate. Both the crystalline phase and the amount of newly formed CaP are strongly dependent on the Ca^{2+} and phosphate concentration in the

Table 3.4 Solubility and pH stability of some biologically relevant calcium phosphates.

Compound	Formula	Solubility $-log\ (K_S)$	pH stability in aqueous solution (25 °C)
α-Tricalcium phosphate (α-TCP)	$\alpha\text{-Ca}_3(PO_4)_2$	25.5	NA[a]
β-Tricalcium phosphate (β-TCP)	$\beta\text{-Ca}_3(PO_4)_2$	28.9	NA[a]
Amorphous CaP (ACP)	$Ca_xH_y(PO_4)_z \cdot nH_2O$ $n = 3\text{--}4.5$	25.7	5–12
Ca-deficient hydro- xyapatite (CDHA)	$Ca_{10-x}(HPO_4)(PO_4)_{6-x}$ $(OH)_{2-x}$ $(0 < x < 1)$	85.1	6.5–9.5
Hydroxyapatite (HA)	$Ca_{10}(PO_4)_6(OH)_2$	116	9.5–12

[a]These compounds cannot be precipitated from aqueous solutions.

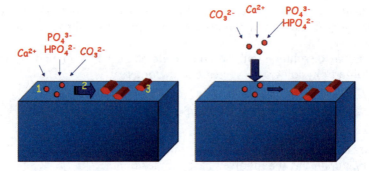

Figure 3.6   Heterogeneous (left) and homogeneous (right) precipitation of CaP nanoceramics.

solutions. This fact points out that not only heterogeneous precipitation takes place at the bioceramic surface, but also homogeneous precipitation must occur during the biomimetic CaP formation (Figure 3.6).

3.4.3.2   *Precipitation of Nanoapatites on CaP Bioceramics*

Whereas dissolution studies are recommended to be carried out in simple buffered solutions such as Tris buffer, biomimetic CaP precipitation reactions can be set off into simulated body fluids with ionic composition similar to that of physiological fluids (Table 3.1). These solutions are highly saturated in phosphates, calcium and carbonates (among other chemical species) and tend to precipitate onto the surface of bioactive ceramics as bone-like apatite phases or CaP precursors, for instance amorphous calcium phosphates (ACP) and octacalcium phosphate (OCP).

The concept of "bone-like apatite" includes the observation that this biomimetic compound shows the apatite crystalline structure, exhibits calcium deficiency, possesses carbonate groups in the unit cell and, from the microstructural point of view, exhibits a small crystallite morphology (often needle-like). It can be said that the biomimetic apatite structure is very similar to the mineral phase of natural bone, although the kind of solution used during the biomimetic process will determine the similarity degree. Determining the nanoapatite precipitation onto CaP bioceramics it is not a trivial issue. In fact, the apatite formation on calcium-phosphate-based ceramics has been the focus of much research for over a decade. However, convincing evidence of apatite identification does not often occur. Sometimes, researchers mainly rely on the diffraction methods to identify crystal structure. It has been a challenge to identify crystal structure of precipitates formed on surfaces of bioceramics because the small quantity of precipitates generates very weak peak intensities in powder XRD analysis. Identifying microcrystals formed on the surfaces of bioactive calcium phosphates is even more difficult because the strong peaks of the substrates overlap the precipitate peaks in the powder XRD pattern.[97] Electron diffraction (ED), transmission electron microscopy (TEM) and Fourier

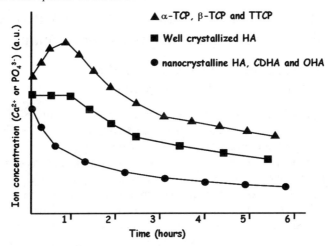

Figure 3.7 Ca^{2+} and PO_4 concentration in SBF *vs.* soaking time of different calcium phosphates.

transform infrared spectroscopy should be considered as more effective tools than powder XRD for identifying precipitation phases formed on bioceramics.

After soaking a CaP bioceramic in SBF, the first measurable parameter is the induction time, *i.e.* the time prior to a detectable decrease in Ca^{2+} and PO_4^{3-} concentrations of the fluid as a result of precipitation[98] (see Figure 3.7). Before this point, the Ca^{2+} and PO_4^{3-} concentrations can remain constant with respect to the initial concentrations, or can increase during induction time. When a decrease is observed from the beginning, it is said that the induction time is equal to zero. As is displayed in Figure 3.7, HA with low crystallinity degree, calcium-deficient HA and oxy-hydroxyapatite shows zero induction time, whereas more soluble CaP such as α-TCP, β-TCP and TTCP lead to an increase of the Ca^{2+} and PO_4^{3-} ion concentrations during the induction times. Well-crystallised HA does not commonly show ionic concentration increase during the induction times.

The ionic concentration of the fluid clearly determines the precipitation or not of biomimetic calcium phosphates. For instance, when HA and β-TCP are immersed into Tris-buffer, or phosphate-containing Homsy's pseudoextracellular fluid (without Ca^{2+}) no new CaP phase is formed on the surfaces.[47] The limited solubility of HA and TCP may be the main reason they failed at surface CaP formation in these solutions. This assumption is supported by the fact that in other biomimetic solutions, initially saturated with Ca^{2+} and phosphate, HA and TCP produced calcium phosphate layers on their surfaces. Thus, in terms of surface change, sufficient concentrations of both Ca^{2+} and phosphate are essential for low-solubility HA and β-TCP. Therefore, under high Ca^{2+} and phosphate concentrations, all the calcium phosphates bioceramics considered so far in this chapter can develop a new phase on the surface. However, kinetics, compositions and structures of the new phases are significantly different.

3.5 Biomimetic Nanoceramics on Hydroxyapatite and Advanced Apatite-Derived Bioceramics

3.5.1 Hydroxyapatite, Oxyhydroxyapatite and Ca-Deficient Hydroxyapatite

Since 1970, the beneficial effects of hydroxyapatite implants have been the subject of study for the biomaterials scientist community. Crystalline hydroxyapatite is a synthetic material analogue to calcium phosphate found in bone and teeth,[99] and a highly cytocompatible material that has been considered for coating on metallic implants,[100] porous ceramic that facilitates bone in-growths,[101] inorganic component in a ceramic-polymer composite,[102] granulate to fill small bone defects[103] and for tissue-engineering scaffolds.[104]

Besides its excellent biocompatibility, synthetic HA mimics many properties of natural bone.[12] HA allows a specific biological response in the tissue-implant interface, which leads to the formation of bonds between the bone and the material.[105] As described above, this response is mediated by solution, precipitation and ionic exchange reactions that result in the surface transformation into a biomimetic nanoceramic formed surface.

The data indicate that the behaviour of the hydroxyapatite family upon immersion in most of the simulated physiological solutions was structure and composition dependent. The structural effect is a combination of crystallinity and specific surface area, since these structural properties varied in parallel.

When apatite is soaked in any Ca^{2+}- and PO_4-containing simulated body fluid, the variations observed within the fluid are essential to understand the biomimetism of these compounds. The crystallinity and dehydroxilation degree, stoichiometry, *etc.* affect the apatite reactivity. In this sense, low-crystalline HA (for instance synthesised by wet methods and treated below 700 °C) incorporates Ca^{2+} and phosphate ions from the solution immediately after coming into contact with the fluid. Therefore, low-crystalline HA exhibits induction time equal to zero. Similar behaviour is shown by Ca-deficient HAp and those highly dehydroxilated HA as well. In fact, the Ca^{2+} and PO_4 incorporation is initially as intense as the precipitation that occurred in super-saturated solutions when CaP seeds are soaked within them. The Ca^{2+} and PO_4 incorporation gradually decreases as the solid/solution equilibrium is reached.

On the contrary, the reaction that takes place onto crystalline HA is significantly different. In the absence of measurable dissolution processes, crystalline HA shows induction times of around 1 hour (at pH 7.4 and 37 °C). From then on, a Ca^{2+} and PO_4 decrease can be measured in the SBF.

In addition to the induction time, the Ca/P molar ratio of the newly formed calcium phosphate provides essential information for elucidating its crystal-chemical characteristics. For instance, Ca/P molar ratios of 1.75–1.79 are commonly calculated for the biomimetic CaP precipitated onto Ca-deficient HAp (CDHA). In this case, the Ca/P ratio is higher than 1.67, which indicates that the newly formed CaP is a type-B carbonate apatite, in which PO_4^{3-} substitutes for CO_3^{2-}. This type of apatite commonly occurs during biological mineralisation processes and consist of solid solutions, whose compositions can vary between $Ca_{10}(PO_4)_6(OH)_2$ and $Ca_8(PO_4)_4(CO_3)_2$. In the case of biomimetic CaP precipitated onto low-crystalline HA, the Ca/P molar ratio is around 1.66, that is very close to stoichiometric HA, whereas those precipitated onto crystalline HA and OHA are 1.34–1.40 and 1.45, respectively, far from the Ca/P ratio of 1.67 or the stoichiometric HA.

As mentioned before, the characterisation of biomimetic CaP by X-ray diffraction is very difficult when they are precipitated onto synthetic apatites. Therefore, techniques such as Fourier transform infrared spectroscopy (FTIR) and transmission electron microscopy (TEM) play a fundamental role in these kinds of studies to follow the biomimetic process on the surface of CaP-based bioceramics. Table 3.5 shows the FTIR characteristic data of the biomimetic evolution for several CaP bioceramics.

Crystalline HA does not exhibit significant changes in the FTIR spectra after being soaked in SBF. The slight formation of an amorphous phase that incorporates a small amount of carbonates is observed. Crystalline HA shows a very slow kinetic for the reactions that constitute the bioactive process (dissolution, precipitation and ionic exchange) and, consequently, several strategies have been proposed to upgrade their biomimetic capacity.

3.5.2 Silicon-Substituted Apatites

The biomimetic behaviour of HA can be improved by introducing some substitutions in the structure.[106] The apatite structure can incorporate a wide

Table 3.5 FTIR absorption bands modifications for several apatites during the first week soaked in SBF.

Bioceramic	FTIR spectra evolution
CDHA	• Appearance/increase at $875\,cm^{-1}$ and $1418–1460\,cm^{-1}$ region of C–O characteristic bands. • Gradual reduction of the splitting of the PO_4^{3-} absorption bands at $600,550\,cm^{-1}$ and $1100–1000\,cm^{-1}$ corresponding to the formation of amorphous or low-crystalline CaP phases.
Nano-HA	• Appearance/increase at $875\,cm^{-1}$ and $1418–1460\,cm^{-1}$ region of C–O characteristic bands. • Gradual reduction of the splitting of the PO_4^{3-} absorption bands at $600,\,550\,cm^{-1}$ and $1100–1000\,cm^{-1}$ corresponding to the formation of amorphous or low-crystalline CaP phases.
Crist-HA	• Gradual reduction of the splitting of the PO_4^{3-} absorption bands at $600,\,550\,cm^{-1}$ and $1100–1000\,cm^{-1}$ corresponding to the formation of amorphous or low-crystalline CaP phases.
OHA	• Appearance/increase at $875\,cm^{-1}$ and $1400–1500\,cm^{-1}$ region of C–O characteristic bands. • Appearance at $632\,cm^{-1}$ corresponding to the librational mode of OH. • Occasionally, appearance at 559 and $525l\,cm^{-1}$ corresponding to octacalcium phosphate formation.

variety of ions, which affect both its cationic and anionic sublattices. For example, in biological apatites CO_3^{2-} by PO_4^{3-} (type B) or OH^- (type A) are likely substitutions.[107,108] In the case of B-type carbonated apatites, the neutrality is usually reached by the incorporation of single-valence cations (Na^+ or K^+) in the Ca^{2+} positions.[109,110]

Studies carried out by Carlisle[111,112] have shown the importance of silicon in bone formation and mineralisation. This author reported detection of silicon *in vivo* within the unmineralised osteoid region (active calcification regions) of the young bone of mice and rats. Silicon levels up to 0.5 wt% were observed in these areas, suggesting that Si has an important role in the bone calcification process. Moreover, the highest bioactivity of silica-based glasses and glass-ceramics (and the mechanism proposed for the bioactive behaviour),[113,114] suggested that the silicon incorporation into apatites would improve the *in vivo* bioactive performance. New apatite layers are formed on the surface of bio-active silica-based glasses and glass-ceramics after a few hours in simulated body fluids. The silanol groups (Si-OH) formation has been proposed as a catalyst of the apatite phase nucleation, and the silicon dissolution rate is considered to have a major role on the kinetics of this process.[115,116] These events suggested the idea of incorporating Si or silicates into the HA structure.

Si-substituted hydroxyapatites (SiHA) are among the most interesting bio-ceramics from the biomimetic point of view. *In vitro* and *in vivo* experiments have evidenced an important improvement of the bioactive behaviour with respect to nonsubstituted apatites.[117,118] Figure 3.8 shows the scanning electron micrographs of pure HA and SiHA after five weeks soaked in SBF. The surface of pure HA remains almost unaltered at the SEM observation since the slow surface reactivity does not allow the observation of significant changes under these conditions. On the contrary, SiHA develops a new apatite phase with a different morphology with respect to the substrate. The surface of SiHA appears covered by a new material with acicular and plate-like morphology, characteristic of new apatite phases grown on bioactive ceramics.

Figure 3.8 SEM micrographs of HA and a silicon-substituted HA after five weeks in SBF.

The term silicon-substituted means that silicon is substituted into the apatite crystal lattice and is not simply added. Silicon or silicates are supposed to substitute for phosphorus, or phosphates, with the subsequent charge unbalance.[119] The amount of silicon that can be incorporated seems to be limited. The literature collects values ranging from 0.1 to 5% by weight in silicon.[120–122] Small amounts of 0.5 and 1% are enough to yield important biomimetic improvements.

The controlled crystallisation method is, by far, the most common synthesis route to obtain SiHA found in the scientific literature.[117–120,123] This process comprises the reaction of a calcium salt or calcium hydroxide with orthophosphoric acid or a salt of orthophosphoric acid in the presence of a silicon-containing compound. Under these conditions it is believed that the silicon-containing compound yields silicon-containing ions, such as silicon ions and/or silicate ions, which substitute in the apatite lattice. There are several synthetic routes to incorporate Si into the hydroxyapatite structure[119,121,124–126] and the kind of silicon precursor, as well as the synthesis method, can lead to different SiHA with different chemical and physical properties. This is clear in the case of the thermal stability of these compounds.

The amount of silicon incorporated also has an important influence on the thermal stability. For instance, when a series of SiHA with nominal formula $Ca_{10}(PO_4)_{6-x}(SiO_4)_x(OH)_{2-x}$, for $x = 0$, 0.25, 0.33, 0.5 and 1 are prepared, using TEOS as silicon source, the as-precipitated samples are always a single nanocrystalline apatite phase (Figure 3.9). After heating at 900 °C, samples with Si content up to 0.33 remained as single apatite phase, whereas higher Si content led to the decomposition changing into hydroxylapatite and α-TCP.[127,128] In fact, this is an appropriated method to obtain biphasic material α-TCP-HA at relatively low temperature. α-TCP is a high-temperature phase that appears when HA or β-TCP is treated over 1200 °C. The presence of silicon seems to stabilise the α-TCP at lower temperatures.

3.5.2.1 *Crystal-Chemical Considerations of SiHA*

Silicon (or SiO_4^{4-}) for P (or PO_4^{3-}) is a nonisoelectronic substitution. This means that the extra negative charge introduced by SiO_4^{4-} must be compensated by means of some mechanism, for example creating new anionic vacancies. The Si, or SiO_4^{4-}, incorporation into the apatite structure at the P, or PO_4^{3-}, position has been studied by several authors. Gibson *et al.*[119] have reported on the structure of aqueous precipitated SiHA. The main structural evidences reported were the decrease and increase of *a* and *c* parameters, respectively, absence of secondary phases and the increase of tetrahedral distortion. These authors have proposed a mechanism to compensate the negative charge introduced by the SiO_4^{4-} incorporation, in apatites obtained by aqueous precipitation method. They show the formation of vacancies at the OH^- site, in a mechanism that can be summarised as follows:

$$PO_4^{3-} + OH^- \leftrightarrow SiO_4^{4-} + \mu$$

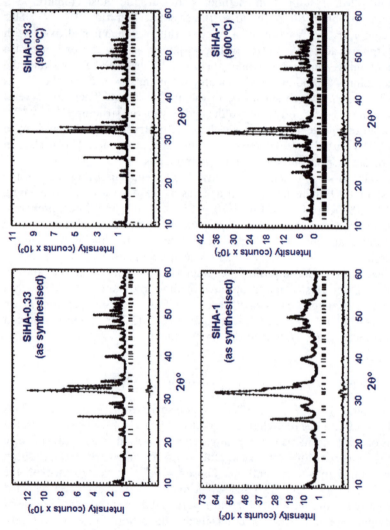

Figure 3.9 Powder X-ray diffraction pattern of $Ca_{10}(PO_4)_{6-x}(SiO_4)_x(OH)_{2-x}$, for $x = 0.33$ and 1. As synthesised samples (left) and treated at 900 °C (right) are showed. The vertical lines mark the positions of Bragg peaks for an apatite-like phase and α-TCP (only for SiHA-1 treated at 900 °C).

obtaining Si-substituted apatites with general formula $Ca_{10}(PO_4)_{6-x}(SiO_4)_x$ $(OH)_{2-x}\mu_x$.

The structural analysis of SiHA has been carried out by X-ray diffraction studies. However, this technique does not allow distinguishing between P and Si, since they are almost isoelectronic, and the presence of H atoms cannot be determined by this technique. Neutron diffraction data seems to be an appropriate method for the structural study of Si-substituted HA.[129–131] In order to explain the higher reactivity of SiHA, the neutron diffraction studies have been focused on the hydrogen atoms in the OH$^-$ groups. This group has great importance in the reactivity of these compounds. As can be seen in Figure 2.13 the thermal displacement of the H atom along the *c*-axis is more than twice that for SiHA. This disorder, together with the tetrahedral distortion resulting from the substitution of PO_4^{3-} by SiO_4^{4-}, could contribute to the higher reactivity of SiHA. However, a crystal-chemical explanation of the SiHA-improved biomimetism would be clearly insufficient. The biomimetic process is a surface process, which is enhanced by the material reactivity. The sum of the different factors may justify the enhanced reactivity. From the point of view of *crystalline structure*, silicon yields tetrahedral distortion and disorder at the hydroxyl site, which could decrease the stability of the apatite structure and, therefore, increase the reactivity. From the point of view of the *microstructural level*, the changes are even more evident. Grain-boundary defects are the starting points of dissolution under *in vivo* conditions. There is a close relationship between the amount of silicon, the number of sintering defects at the grain boundaries and the dissolution rate. In particular, the number of triple junctions in SiHA may have an important role in the material reactivity and consequently, in the rate at which the ceramic reacts with the bone. Finally, the *surface charge* undergone by the ceramic due to the presence of SiO_4^{4-} would also play an important role for the Ca^{2+} incorporation at the new biomimetic layer. This effect could also be responsible in part for the alteration in its biological response. Summarising, the understanding of the improved biomimetic behaviour in SiHA requires to be considered as a sum of different factors at different levels.

3.6 Biphasic Calcium Phosphates (BCPs)

3.6.1 An Introduction to BCPs

Nowadays, the general requirements for ideal implants aimed at bone regeneration establish that they should exhibit pores of several hundred micrometres, a biodegradation rate comparable to the formation of bone tissue (*i.e.* between a few months and about 2 years) and sufficient mechanical stability.[3] HA and TCP (both, α and β polymorphs) do not fulfil these requirements and some clinical failures have occurred as a consequence of inappropriate biodegradability kinetics, which eventually will involve a disadvantage to the host tissue surrounding the implant. For instance, some implants made of calcined

HA to reconstruct mandibular ridge defects have resulted in high failure rate in human clinical applications.[132] In order to avoid this problem, the use of granular instead of block forms of HA was suggested,[133] although HA exhibit some drawbacks due to lack of biodegradability, independently of the implant form. On the other hand, β-TCP ceramics have been developed as a biodegradable bone replacement and commercially available as, for instance, ChronOS™, Vitoss™, *etc.*[134] However, when used as a biomaterial for bone replacement, the rate of biodegradation of TCP has been shown to be too fast. In 1988, Daculsi *et al.*[83] thought that the presence of an optimum balance of stable HA and more soluble β-TCP should be more favourable than pure HA and β-TCP. Due to the biodegradability of β-TCP component, the reactivity increases with the β-TCP/HA ratio. Therefore, the bioreactivity of these compounds could be controlled through the phase composition. The main advantage with respect to other nonsoluble calcium phosphates is that the mixture is gradually dissolved in the human body, acting as a stem for newly formed bone and releasing Ca^{2+} and PO_4^{3-} to the local environment.[135] *In vivo* tests have confirmed the excellent behaviour of BCP (biphasic calcium phosphate) concerning the biodegradability rate.[136–139]

Since Ellinger *et al.*[136] termed for the first time a CaP as BCP, to describe a mixture of β-TCP and HA, many advances have occurred in the BCP field. The works carried out by Daculsi *et al.*[138,140] impelled the commercialisation of BCP and currently can be found as trademarks like Triosite™, HATRIC™, Tribone™, *etc.* Nowadays, BCPs are clinically used as an alternative or as an additive to autogenous bone for dental and orthopaedic applications. Implants shaped as particles, dense or porous blocks, customised pieces and injectable polymer-BCP mixtures are common BCP-based medical devices. Moreover, research is in progress to enlarge the clinical applications to field of scaffolding for tissue engineering[141–143] and carriers loading biotech products.[144,145]

HA chemistry and structure have been widely explained in Chapters 1 and 2. β-TCP is a phase that crystallises in the rhombohedral system, with a unit cell described by the space group R3Ch and unit cell parameters $a = 10.41$ Å, $c = 37.35$ Å, $\gamma = 120°$. At temperatures above 1125 °C it transforms into the high-temperature phase α-TCP. Being the stable phase at room temperature, β-TCP is less soluble in water than α-TCP. Pure β-TCP never occurs in biological calcifications, *i.e.* there is no biomimetic process that result in β-TCP. Only the Mg-substituted form (withlockite) is found in some pathological calcifications (dental calculi, urinary stones, dentinal caries, *etc.*).

α-TCP is usually prepared from β-TCP by heating above 1125 °C and it might be considered as a high-temperature phase of β-TCP. α-TCP crystallises in the monoclinic system, with a unit cell described by the space group P2₁/a and unit cell parameters of $a = 12.89$ Å, $b = 27.28$ Å, $c = 15.21$ Å, $\beta = 126.2°$. Therefore, α-TCP and β-TCP have exactly the same chemical composition but they differ in their crystal structure. This structural difference determines that β-TCP is more stable than the α-phase. Actually, α-TCP is more reactive in aqueous systems; has a higher specific energy and it can be hydrolysed to a mixture of other calcium phosphates. Similarly to the β-phase, α-TCP never occurs in biological calcifications, and it is occasionally used in calcium phosphate cements.[86,146,147] In recent years, α-TCP is being used as a component of biphasic HA- α-TCP bioresorbable scaffolds. This material is obtained by heating silicon-substituted HA at temperatures around 1000 °C, obtaining the so-named silicon stabilised α-TCP.[84,148–150]

3.6.2 Biomimetic Nanoceramics on BCP Biomaterials

As described above, the biomimetic process in calcium phosphates is based on dissolution, precipitation and ion-exchange processes. The dissolution rate of BCPs depends on the ratio of TCP to HA in the compound.[151,152] Under *in vivo* conditions, the calcium phosphate ceramics containing a greater amount of TCP phase also show greater biodegradation. Many factors influence both dissolution and biodegradation, including the size and the conditions under which HA and TCP are synthesised. Interfacial aspects include stability when subjected to body fluid, porosity of surface and grain-boundary condition. However, the most important factor determining the dissolution and biodegradability is the TCP to HA ratio.[153] At a pH range of 4.2–8.0 and therefore at the physiological pH 7.4, HA is less soluble than other tricalcium phosphates. In fact, the tricalcium phosphate dissolves 12.3 times faster than HA in acidic medium and 22.3 times faster than HA in basic medium.

When β-TCP/HA biphasic materials are soaked into a simulated physiological solution, SBF for instance, the pH values of any experimental solution decreases to values between 4.6 and 6.0 after several weeks of immersion. This fact is consistent with most of the biomimetic processes, which evidences a pH decrease of the solution during the calcium phosphate precipitation. HA does not dissolve, but β-TCP is subject to dissolution. The interaction of the TCP phase with the solution takes place in a very short time after soaking.[153]

Following the phase content by XRD patterns collected as a function of soaking time, it can be seen that depending on the TCP amount contained in the BCP, from 25 up to 100% of the β-TCP or α-TCP contained can be dissolved after 4 weeks of soaking. However, BCP does not only degrade under the action of physiological solutions. In fact, the changes in weight of most BCP materials tested after immersion in SBF are negligible, which means that the precipitation of new CaP phases (biomimetic ones) and/or hydrolysis of the TCP phase also take place on the surface of the materials.

Whereas the dissolution process seems to be an easy question to resolve in BCPs, the precipitation process becomes more difficult to understand. In fact, the newly formed phases are different when the biomimetic process is carried out under static or dynamic conditions and, of course, completely different for *in vivo* experiments.[97] In static biomimetic conditions, calcium and phosphate ions from the TCP phase in the BCP dissolves into solution and reprecipitates on to the BCP as calcium-deficient HA. Due to its greater stability, the HA of the BCP acts as a seed material in physiological solutions. However, a deep study of the biomimetic CaP formed under dynamic conditions can show a different scenario for the BCPs. Single-crystalline precipitates of calcium phosphates on porous BCP bioceramics obtained after immersion in dynamic simulated body fluid (SBF) and after implantation in pig muscle were examined using electron diffraction in a transmission electron microscope. The crystals formed *in vitro* in dynamic SBF were identified as octacalcium phosphate (OCP), instead of apatite. The hard evidence provided by single-crystal diffraction indicates that the precipitation on BCP in SBF may be neither "bone-like" nor "apatite".

3.7 Biomimetic Nanoceramics on Bioactive Glasses

3.7.1 An Introduction to Bioactive Glasses

Bioactive glasses were discovered by Prof. L.L. Hench in 1971 and nowadays are considered as the first expression of bioactive ceramics. Due to the high bioactivity level and their brittleness, these materials find clinical application in those cases where high tissue regeneration is required without supporting high loads or stresses. Currently, they are used for replacement of ear bones and, as powders for periodontal surgery and bone repairing.[56]

The starting point for the first bioglass synthesis was based upon the following simple hypothesis:[154]

"The human body rejects metallic and synthetic polymeric materials by forming scar tissue because living tissues are not composed of such materials. Bone contains a hydrated calcium phosphate component, hydroxyapatite and therefore if a material is able to form a HA layer in vivo it may not be rejected by the body"

Actually, the apatite phase formed on the surface of bioactive glasses is calcium-deficient, carbonate-containing, nanocrystalline and therefore very similar to the biological ones. The first *in vivo* experiments carried out with the so-named 45S5 Bioglass® (see Figure 3.10) demonstrated that these apatite crystals were bonded to layers of collagen fibrils produced at the interface by osteoblasts. This chemical interaction between the newly formed apatite layer and the collagen fibrils constitutes a strong chemical bond denoted a "bioactive bond".[155,156]

Bioactive glasses exhibit Class A bioactivity, *i.e.* they are *osteoproductive*[i] materials,[157] instead of those ceramics such as HA that behave as *osteoconductive*[ii] materials and are classified as Class B bioactive materials. Since both kinds of materials are bioactive, they form a mechanically strong bond to bone. However, as a Class A bioactive material, bioactive glasses exhibit a higher rate of bonding to hard tissues (although they also bond to soft tissues).

---

[i] Osteoproduction is the process whereby a bioactive surface is colonised by osteogenic stem cells free in the bone defect environment as a result of surgical intervention.

[ii] Osteoconduction is the process of bone migration along a biocompatible surface.

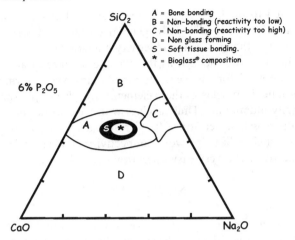

Figure 3.10 Compositional diagram for the bone-bonding ability of melt-derived glasses.

Together with a rapid bonding to bone, bioactive glasses also show enhanced proliferation compared to calcium phosphate bioceramics or any other Class B bioactive ceramic.

3.7.2   Composition and Structure of Melt-Derived Bioactive Glasses

The first bioactive glass reported in 1971 was synthesised in the system SiO_2-P_2O_5-CaO-Na_2O.[7] The glass composition of 45% SiO_2 – 24.5% Na_2O – 24.5% CaO – 6% P_2O_5 was selected. This composition provides a large amount of CaO with some P_2O_5 in a Na_2O-SiO_2 matrix and it was very close to a ternary eutectic and, therefore, easy to melt. Actually, the synthesis process consisted of melting the precursor mixture and quenching. In the following years, several compositions contained in the phase equilibrium diagram of such systems were studied.[158–160]

Silicate glasses can be considered as inorganic polymers, whose monomer units are SiO_4 tetrahedra. These units are linked through the O placed at the tetrahedral apexes (bonding oxygen atoms) and the polymeric network is disrupted when the oxygen atoms are not shared with another SiO_4 tetrahedron (nonbonding oxygen). The presence of cations such as Na^+ and Ca^{2+} in the bioglass composition causes a discontinuity of the glassy network through the disruption of some Si–O–Si bonds. As a consequence, nonbridging oxygens are created. The network modifiers are in this case MO and M_2O-type oxides like CaO and Na_2O, respectively. The properties of such glasses may be explained on the basis of the crosslink density of the glass network using concepts taken from polymer science that are normally used to predict the behaviour of organic polymers. The network connectivity (NC) or the crosslink density of a glass can be used to predict its surface reactivity and solubility among other physical-chemical properties.[161,162] In general terms, the lower the crosslink density of the glass, the greater the reactivity and solubility.

The crosslink density is defined as the average number of additional cross-linking bonds above 2 for the elements other than oxygen forming the glass network backbone. In bioglasses these elements are silicon, phosphorus, boron and occasionally aluminium. Thus, a glass with a network connectivity of 2, equivalent to a crosslink density of 0, correspond to a linear polymer chain, while a pure silica glass has a network connectivity of 4. The calculation of the network connectivity is a very easy operation defined by:

$$NC = 8 - 2R \qquad (3.3)$$

R = number of O atoms/number of network-former atoms (Si, P, B or Al).

New components were added to the system almost simultaneously in order to act as network formers and/or modifiers and to decrease the synthesis temperature of bioglasses. But the main purpose of their inclusion was to improve their properties focused on clinical applications, *i.e.* to increase their bioreactivity or at least to preserve or increase their bioactivity, while adding new properties to the materials. In this sense, the addition of K_2O, MgO, CaF_2, Al_2O_3, B_2O_3 or Fe_2O_3 were tested.[163] But all these efforts did not always lead to positive results, since the addition of some of these oxides degraded or totally avoided the bioactive behaviour of Bioglass. For instance, a 3% of Al_2O_3 added to the initial composition of Hench, in order to improve its mechanical properties,[164] eliminated its bioactivity, or the addition of Fe_2O_3 to obtain glass-ceramics for hyperthermia treatment of cancer[165] decreased the bioactivity. In 1997, Brink *et al.*[166,167] studied the *in vivo* bone-bonding ability of 26 melt glasses in the system Na_2O-K_2O-CaO-MgO-B_2O_3-P_2O_5-SiO_2, concluding that the compositional limits for bioactivity were: 14–30 mol% of alkali oxides ($Na_2O + K_2O$), 14–30 mol% of alkaline earth oxides (CaO + MgO), and less than 59 mol% of SiO_2.

3.7.3 Sol-Gel Bioactive Glasses

The sol-gel method is a synthesis strategy that consists of obtaining a sol by hydrolysis and condensation of the precursors, commonly metal alkoxides and inorganic salts, and the subsequent gelation of the sol. The sol-gel method presents some advantages with respect to melting for glass processing. Glasses are obtained with a higher degree of purity and homogeneity. However, the real potential of sol-gel bioactive glasses is based in two aspects:

a) Sol-gel synthesis offers a potential processing method for molecular and textural tailoring of the biological behaviour of bioactive materials. The inherent features of this method allow obtaining bioactive compositions in the form of particles, fibres, foams, porous scaffolds, coatings and, of course, monoliths.

b) The sol-gel method of glass processing provides materials with high mesoporosity and high surface area, enhancing the kinetics of the apatite formation and expanding the composition range for which these materials show bioactive behaviour.[168-171] It must be taken into account that the reactions that trigger the bioactive behaviour take place on the surface. Therefore not only chemical composition but also the textural properties (pore size and shape, pore volume, *etc.*) play a fundamental role in the development of the biomimetic CHA layer.[172-177]

Among the different bioactive sol-gel glasses, $SiO_2 \cdot CaO \cdot P_2O_5$ is the most widely studied system.[178-181] Each component contributes to the structure–reactivity relationship, so providing the different *in vivo* response for each composition. Silica is a network former and constitutes the basic component of the glasses. Higher amounts of SiO_2 result in more stable glasses. CaO is a network modifier, *i.e.* its presence partially avoids the Si–O–Si link formation, resulting in more reactive glasses. As a general trend, the higher the CaO content, the higher the bioactive behaviour of the glass. Finally, the role of P_2O_5 is not clear from the structural point of view. It can be found as tetrahedral units that contribute to the network formation, or as orthophosphates grouped into clusters. The presence of P_2O_5 contributes to the CHA crystallisation on the glass surface during the bioactive process, although amounts higher than 12% in weight inhibit the bioactivity.

The sol-gel method makes it possible to expand the bioactive compositional range studied in the phase equilibrium diagram of melted glasses, and the glasses so obtained exhibit higher surface area and porosity values, critical factors in their bioactivity.[182,183] This feature allows the simplification of the chemical systems thus obtaining bioactive compositions in the diagram SiO_2-CaO. This binary system was tested by Kokubo and coworkers[184,185] producing glasses of the binary system SiO_2-CaO with a SiO_2 content less than or equal to 65%, prepared by melting; Vallet-Regí and coworkers[61,170,171] prepared also glasses in this system, with SiO_2 contents of up to 90% (50–90% SiO_2), prepared by the sol-gel technique.

The application of the sol-gel chemistry to the synthesis of bioactive glasses, opened new perspectives in the chemistry of these compounds. For the same silica content, the rate of CHA formation is higher in sol-gel-derived glasses than in melt-derived ones. The higher bioactivity of the sol-gel glasses is attributed to the high surface area and concentration of silanol groups on the surface of these materials. These features come from the sol-gel processing that allows production of glasses and ceramics at much lower temperatures compared with conventional methods.

3.7.4 The Bioactive Process in SiO_2-Based Glasses

The cascade of events that leads to the growth of a nanoapatite phase and the subsequent bonding between glass and bone has been described by Hench *et al.*[186] The basis for bone bonding is the reaction of the glass with the surrounding solution. A sequence of interfacial reactions, which begin immediately after the bioactive material is implanted, leads to the formation of a CHA layer and the establishment of an interfacial bonding. Hench summarises the sequence of interfacial reactions as follows:

1) Rapid exchange of Na^+ or Ca^{2+} with H^+ or H_3O^+ from solution and formation of silanols (Si-OH) at the glass surface.

$$Si\text{-}O\text{-}Na^+ + H^+ + OH^- \rightarrow Si\text{-}OH + Na^+ + OH^-$$

2) Loss of soluble silica, in the form of $Si(OH)_4$ resulting from breaking of Si–O–Si bonds and formation of silanols.

$$2(Si\text{-}O\text{-}Si) + 2(OH^-) \rightarrow Si\text{-}OH + OH\text{-}Si$$

3) Condensation of silanols to form a hydrated silica gel layer.

$$2(Si\text{-}OH) + 2(OH\text{-}Si) \rightarrow \text{-}Si\text{-}O\text{-}Si\text{-}O\text{-}Si\text{-}O\text{-}Si\text{-}O\text{-}$$

4) Migration of Ca^{2+} and PO_4^{3-} groups to the surface through the silica layer, forming a $CaO\text{-}P_2O_5$-rich film on the top of the silica-richer layer.
5) Crystallisation of the amorphous calcium phosphate layer by incorporation of OH^-, CO_3^{2-}, or F^- from solution to form a mixed hydroxyl-carbonate apatite layer (CHA) or hydroxyl-carbonate fluorapatite (HCFA) layer from the solution.
6) Adsorption of biological moieties in the HCA layer.
7) Action of the macrophages.
8) Attachment of the stem cells.
9) Differentiation of the stem cells.
10) Generation of the collagen matrix.
11) Crystallisation of the mineral matrix.

Stages 1 to 5 occur under *in vitro* conditions and do not require any biological or organic entity. Therefore, these stages constitute the mechanism that rules the synthesis of biomimetic apatites on bioactive glasses.

3.7.5 Biomimetic Nanoapatite Formation on SiO_2-Based Bioactive Glasses: The Glass Surface

The problem of the mechanism of apatite formation on the surfaces of glasses and glass-ceramics was a controversial topic during the 1990s. The body fluid and artificial SBFs are supersaturated with respect to the apatite under the normal condition. Under such an environment, once the apatite nuclei are formed on the surfaces of glasses and glass-ceramics, they can grow spontaneously by consuming the calcium and phosphate ions from the surrounding solution. The problem is therefore reduced to the mechanism of the apatite nucleation on the surfaces of glasses.

SiO_2-CaO- and SiO_2-CaO-Na_2O-based glasses, including those with and without P_2O_5, form the apatite layer on their surface *in vivo* as well as *in vitro* by biomimetic processes. On the contrary, CaO-P_2O_5-based glasses do not develop such phases,[187] indicating that the SiO_2 presence is mandatory to set off the bioactive process. Calcium ions dissolve from the glass and increase the degree of the supersaturation of the surrounding body fluid with respect to the apatite, and the hydrated silicate ion formed on their surfaces might provide favourable sites for the apatite nucleation. The importance of the hydrated silicate ion in forming the apatite layer had been also proposed by Hench, as mentioned above.[188,189]

SiO_2-CaO glasses containing a small amount of P_2O_5, for example SiO_2 50-CaO 45- P_2O_5 5 (mol%), develop an apatite layer on their surface in SBF faster (around 6 h) than those compositions without P_2O_5 (around 3 days). These glasses succeeded in developing biomimetic nanoapatites, contrarily to CaO-P_2O_5-based glasses that do not form them.

Since the body fluid is already supersaturated with respect to the apatite under normal conditions, once the apatite nuclei are formed; they can grow spontaneously by consuming the calcium and phosphate ions from the surrounding body fluid. In view of these factors, Ohtsuki *et al.*[114] established that the rate of apatite nucleation on glasses in SBF increases in the order

$$CaO\text{-}P_2O_5 < \, < SiO_2\text{-}CaO < SiO_2\text{-}CaO\text{-}P_2O_5$$

The rate, *I*, of nucleation of a crystal on a substrate in a solution at the temperature, *T*, is generally given by:[190]

$$I = I_0 \exp\left(\frac{-\Delta G^*}{kT}\right) \exp\left(\frac{-\Delta G_m}{kT}\right) \tag{3.4}$$

where ΔG^* is the free energy for formation of an embryo of critical size, ΔG_m is the activation energy for transport across the nucleus/solution interface. Among them, ΔG_m is independent of the substrate. ΔG^* is given by:

$$\Delta G^* = \frac{16\sigma^3 f(\theta)}{3\left(kT/V_\beta \ln\left(IP/K_0\right)\right)^2} \tag{3.5}$$

where σ is interface energy between the nucleus and the solution, *IP* is ionic activity product of the crystal in the solution, K_0 is the value of *IP* at equilibrium, *i.e.* the solubility product of the crystal; $f(\theta)$ is a function of contact angle between the nucleus and the substrate, and V_β is the molecular volume of the crystal phase. Among them, $f(\theta)$ depends upon the substrate, and IP/K_0 is a measure of the degree of supersaturation, which also depends upon the substrate when the substrate releases some constituent ions of the crystal, while others are independent of the substrate.

Experimental results have demonstrated that SiO_2-CaO based glasses dissolve significant amounts of calcium ions, whereas CaO-P_2O_5-based glasses dissolve important amounts of phosphate ions. Consequently, the changes of *IP* in the SBF for both cases are very similar and, therefore, the different biomimetic behaviour cannot be attributed to the larger increase in the degree of the supersaturation due to the dissolution of the calcium ion.

The term $f(\theta)$, generally given by eqn (3.6) decreases with decreasing interface energy between the crystal and the substrate:

$$f(\theta) = \frac{(2 + \cos\theta)(1 - \cos\theta)^2}{4} \tag{3.6}$$

This indicates that the SiO_2-CaO-based glasses provide a specific surface with lower interface energy against the apatite. Bioactive glasses form a silica hydrogel layer prior to the formation of the apatite layer. This layer is responsible of the decrease of $f(\theta)$, decreasing the contact angle and providing specific favourable sites for apatite nucleation.

The studies carried out by Li *et al.*[168] on bioactive sol-gel glasses showed the importance of surface area and porosity in the formation of biomimetic nanoapatites. The apatite growth in SBF was demonstrated for sol-gel glass composition with nearly 90% of SiO_2. The rate of surface HCA formation for 58S composition (see Table 3.6) was even more rapid than for melt-derived 45S5 Bioglass. Table 3.6 shows some of the more often tested compositions with their corresponding nomenclature. More information about the numerous sol-gel glasses compositions can be found in reference 191.

High surface area seems to be very important for SiO_2-based bioactive glasses, both melt-derived and sol-gel glasses. Melt derived glasses initially exhibit surface area values below $1\,m^2\,g^{-1}$. However, they develop more than $100\,m^2\,g^{-1}$ when they come into contact with fluids at physiological pH, as was demonstrated by Greenspan *et al.*[192] Once this surface area is developed, the melt derived bioglasses are suitable to be coated by biomimetic nanoapatites (Figure 3.11).

Table 3.6 Chemical composition (wt%) for some melt derived and sol-gel glasses. [(+)] bioactive glasses; [(−)] nonbioactive glasses.

	SiO_2	P_2O_5	CaO	Na_2O
45S5 melt[(+)]	45	6	24.5	24.5
60S melt[(−)]	60	6	17	17
58S sol-gel[(+)]	48	9	33	–
68S sol-gel[(+)]	68	9	23	–
77S sol-gel[(+)]	77	9	14	–
91S sol-gel[(−)]	91	9	–	–

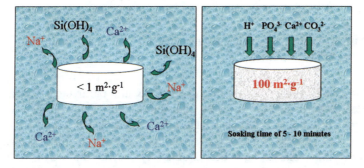

Figure 3.11   Ionic exchange and surface area evolution in bioactive melt-derived glasses after being soaked in SBF.

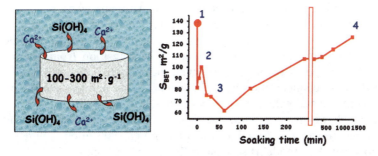

Figure 3.12   Evolution of S_{BET} as a function of soaking time for 58S sol-gel glass.

In the case of sol-gel glasses, the surface evolution is very different compared with, for instance 45S5 melt-derived bioglass.[193] Figure 3.12 shows the S_{BET} evolution of the glass as a function of the soaking time in SBF. Four stages can be clearly differentiated. During the first minute, 1st stage, the glass undergoes a drastic surface decrease from $138\,m^2\,g^{-1}$ (original value) to $82\,m^2\,g^{-1}$, which means a 40% surface reduction in a very short time. This is a very different behaviour compared to melt-derived glasses, which have a very low surface area but develop surfaces of about $100\,m^2\,g^{-1}$ after being soaked in physiological simulated solutions. Afterwards, a partial surface recovering occurs between 1 min and 10 min, 2nd stage, reaching a surface value of $100\,m^2\,g^{-1}$. From this point the glass begins to lose surface gradually, 3rd stage, and after one hour it has lost about the 55% of the initial surface, showing values of $62\,m^2\,g^{-1}$. Finally, the 4th stage involves the progressive surface area recovering from 1 h until the end of the experiment, reaching values of $127\,m^2\,g^{-1}$ after 24 h in SBF. These four stages can be explained in terms of the bioactivity theory of glasses:

a) Loss of surface area due to the fast Ca^{2+} release.

b) Partial surface area restoring due to the Si-OH formation and CO_3^{2-} incorporation.

c) Second surface area loss as a consequence of the amorphous CaP formation.

d) Surface area restoring during the CaP crystallisation into hydroxycarbonate apatite.

In the case of sol-gel glasses, the values of textural parameters depend on the chemical composition of the glass and stabilisation temperature used.[194,195] Moreover, the changes of surface area and porosity depend on the kinetics of the bioactive process for each glass composition. The textural properties of SiO_2-CaO–P_2O_5–glasses have been studied by varying the SiO_2/CaO ratio.[196] This systematic study allowed confirmation that higher presence of SiO_2 results in higher surface area, whereas higher CaO content provides more mesopore volume and larger pore diameter. The morphology of mesopores is modified as a function of SiO_2 (or CaO) content. While the glasses with larger SiO_2 content (80% and 75% mol) have inkbottle-type pores with narrow necks, glasses with lower SiO_2 content (58%, 60%, 65% mol) have cylindrical pores open at both ends with occasional necks along the pores. The pore morphology parallels the variations of pore diameter and volume. The transition from narrow-neck inkbottle-type pores to open-ended cylindrical pores apparently takes place when the pore diameter increases. The higher Ca content leads to the increase of the pore size and volume and causes a change of morphology from inkbottle pores to cylindrical ones. Since the higher ionic concentration occurs into the mesopores, the apatite growth (nucleation and crystallisation) will depend on this porosity. This model is schematically plotted in Figure 3.13.

Although the influence of the texture of the substrate on the formation of apatite is generally admitted, the detailed nature of the nucleation process of the apatite is still a matter of debate. Practically all authors focused the discussion on apatite nucleation upon the role of the silanol groups existing on the glass surface under the environmental conditions where the assays are conducted.[197–199] Wang and Chaki[200] show an epitaxial relationship between

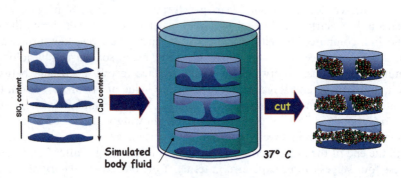

Figure 3.13   Schematic model of the mesopore morphology as a function of SiO_2:CaO ratio. The figure also shows a scheme of apatite formation within the mesopores after soaking in SBF. For the sake of clarity, the apatite layer grown all over the particle free surface is not shown.

Si(111) and apatite in [102] orientation. Interestingly, the phosphorus and calcium of the substrates are generally considered as a mere reservoir that influences the supersaturation of the solution as they are leached from the glass. Nevertheless, phosphorus and calcium as components of bioactive glasses could in fact be potential nucleation centres for apatite crystallisation, although the role of P_2O_5 is controversial as will be explained in the next section.

3.7.6 Role of P_2O_5 in the Surface Properties and the *In Vitro* Bioactivity of Sol-Gel Glasses

In the early 1990s, when the first bioactive sol-gel glasses were prepared in the $CaO–P_2O_5–SiO_2$ system, diverse studies were performed to understand the role of the gel glass constituents in the surface properties and the *in vitro* formation of a CHA phase. In this way, the role of SiO_2 and CaO was reported, but the effect of P_2O_5 was not fully understood.

The bioactive behaviour of $CaO–SiO_2$ glasses demonstrates that P_2O_5 is not an essential requirement for bioactivity, even for high SiO_2 contents.[171] However, even if not essential, P_2O_5 plays an important role on the kinetic formation and final features of the biomimetic apatite growth on glass surfaces. Two series of $CaO–P_2O_5–SiO_2$ glasses were prepared, first with SiO_2 constant (80%),[201] the second with CaO constant (25%) (in mol%).[202] Finally, the nanostructural characterisation of glasses by high-resolution electron microscopy, HRTEM,[203] allows the determination of calcium and phosphorus location in the silica network.

Regarding the *in vitro* bioactivity, it was concluded that P_2O_5 retards the initial *in vitro* reactivity of glasses, defined as the time required for the formation of a layer of amorphous calcium phosphate. However, once some nuclei are formed, for contents of P_2O_5 up to 5%, the growth of CHA crystals in the layer is quicker and yields larger crystals. With respect to the textural characterisation, it was shown that the surface area increases and the diameter and volume of pores decrease when increasing the P_2O_5 content in glasses with 25% of CaO, pointing out that P_2O_5 bonds to CaO, given that increasing the P_2O_5 content produces similar textural effects as decreasing the CaO content.

This assumption has been confirmed by HRTEM since distances between the $[SiO_4^{4-}]$ tetrahedra of 0.53 nm were found in a P-free glass of composition SiO_2 80–CaO 20, in mol%, but only of 0.36 nm were measured in a P-containing glass (SiO_2 80–CaO 17–P_2O_5 3), indicating that in the latter the calcium was out of the glass network. In addition, in P-containing glasses small crystalline clusters (size lower than 10 nm), identified as silicon-doped calcium phosphate nuclei were detected (Figure 3.14).

In P-free glasses bioactivity is controlled by the rapid exchange of calcium in the glass network by protons in solution forming silanol (Si–OH) groups, which attract calcium and phosphorous in SBF to form an amorphous calcium phosphate. Afterwards, a relatively long period is required for the *in vitro* crystallisation of CHA. However, for P-containing glasses the silanol

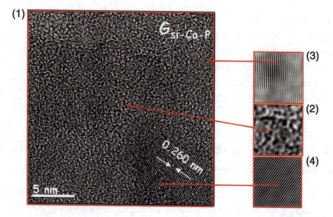

Figure 3.14 Electron microscopy study of a gel glass of composition 17%CaO–3%P_2O_5–80%SiO_2. (1) HRTEM image and (2) filtered HRTEM image of the amorphous matrix. (3 and 4) P–rich crystalline areas oriented along different directions with interplanar spacings close to 0.26 nm.

concentration is lower, retarding the amorphous calcium phosphate formation, but the presence of the mentioned nanocrystals that could act as nucleation centres increase the CHA crystallisation rate.

3.7.7 Highly Ordered Mesoporous Bioactive Glasses (MBG)

By comparing the bioactive behaviour of melt-derived glasses with that of sol-gel glasses, it is easy to understand that increasing the surface area and pore volume may improve the CHA growth on their surfaces. For this reason, ordered mesoporous silica-based materials were proposed for biomimetic purposes. Silica-based mesoporous materials are ordered porous structures of SiO_2, characterised for having high pore volume, narrow pore size distribution and high surface area. SMMs are synthesised by self-assembly of silica-surfactant composites, in which inorganic species (silica precursors) simultaneously condense giving rise to mesoscopically ordered composites formation.[204–207] After removing the surfactant, a silica based mesostructured solid with the textural properties described above is formed (Figure 3.15).

The research group of Prof. Vallet-Regí proposed for the first time the possibility of using silica based mesoporous materials for bone-regenerative purposes. They demonstrated that under specific conditions some structures could develop biomimetic apatites onto the surface. However, high surface areas and porosities are not enough to achieve satisfactory biomimetic behaviour. For instance, MCM-41 is not bioactive and requires to be doped to show bioactivity.[208,209] Other phases like MCM-48 or SBA-15 must be soaked in SBF for 60 and 30 days before developing an apatite-like phase[210,211] Only SBA-15 obtained as coating shows bioactivity after one week, which can be considered reasonable for clinical applications.[212] Although the high-ordered

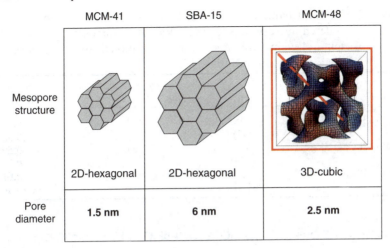

Figure 3.15 Different mesoporous structures.

porosity means an added value over conventional bioactive sol-gel glasses, none of the mesoporous materials described till now improve the bioactive behaviour of the conventional sol-gel glasses.

The real challenge was to obtain bioactive multicomponent sol-gel glasses, with the textural properties of the ordered mesoporous silica. However, a multicomponent glass system is quite complex and consists of mainly amorphous oxides. In 2004, Yan *et al.*[213] demonstrated that the synthesis of highly ordered mesoporous bioactive glasses (MBGs) was possible by templating with a block copolymer. These authors carried out the synthesis of SiO_2-CaO-P_2O_5 ordered mesoporous glasses through the evaporation-induced self-assembly (EISA) method[214] in the presence of a nonionic triblock copolymer ($EO_{20}PO_{70}EO_{20}$), resulting in hexagonal *p6mm* structures. The final materials showed higher biomimetism than bioactive sol-gel glasses obtained by the conventional sol-gel method.

As a consequence of the presence of CaO and P_2O_5 together with the excellent textural properties (see Table 3.7), these MBGs develop a CHA after 4 h in SBF, showing the highest bioactive rate observed up to now for SiO_2-CaO-P_2O_5 systems. Further studies have demonstrated that MBGs are more homogeneous in composition compared to conventional bioactive glasses.[215] Since the inorganic species are distributed homogeneously in the silica network at the nanoscale level (the wall thickness of MBG is <7 nm) these species do not aggregate or become heterogeneous even when the structure density is increased at high calcination temperatures. Secondly, MBGs with different compositions basically exist in the form of a noncrystalline state, in contrast to conventional sol-gel-derived BGs, which frequently show calcium-phosphate-rich clusters due to chemical inhomogeneities.

These materials are excellent as materials for bone grafting and subsequent resorption. After soaking MBGs in water for 2.5 days, the mass losses for Ca,

Table 3.7 Textural parameters obtained by N_2 adsorption porosimetry for ordered mesoporous glasses. Values in brackets correspond to textural values obtained for conventional sol-gel glasses with analogous compositions.

MBG composition (% mol)	S_{BET} ($m^2 g^{-1}$)	Average pore diameter (nm)	Pore volume ($cm^3 g^{-1}$)
58 SiO$_2$-37CaO-5P$_2$O$_5$	195 (95)	9.45	0.46 (0.35)
75 SiO$_2$-20CaO-5P$_2$O$_5$	393 (175)	6.0	0.59 (0.21)
85 SiO$_2$-10CaO-5P$_2$O$_5$	427 (227)	5.73	0.61 (0.24)

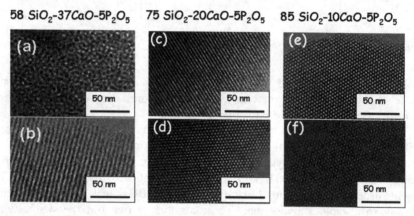

Figure 3.16 TEM images corresponding to $SiO_2 \cdot CaO \cdot P_2O_5$ mesoporous bioactive glasses with different CaO contents: Images (a) nonordered worm-like structure and (b) hexagonal structure correspond to a material with 37% in mol of CaO added during the synthesis. Images (c) hexagonal structure and (d) orthorhombic structure correspond to a material with 20% in mol of CaO added during the synthesis. Images (e) and (f) cubic structure, correspond to a material with 10% in mol of CaO added during the synthesis.

P, and Si species are around 35%, 6%, and 48%, respectively, suggesting that MBGs may have excellent degradability in body fluid, which is important for prospective bioapplications.

In the case of multicomponent systems, such as $SiO_2 \cdot CaO \cdot P_2O_5$, not only the surfactant amount but also the CaO content determine the structure of the MBGs.[216]

Figure 3.16 collects the TEM images for three different compositions, which only differ in the amount of CaO and correspond to the composition displayed in Table 3.7.

A progressive evolution from 2D-hexagonal to cubic structures is observed when decreasing the CaO content. These structural modifications can be explained in terms of the influence of the Ca^{2+} cations on the silica condensation. Ca^{2+} cations act as network modifiers, decreasing the network connectivity. Consequently, the inorganic/organic volume ratio of the micelle is increased with the Ca^{2+} content, thus inducing the formation of hexagonal phases rather than cubic ones.

As a general rule in "conventional" SiO_2-CaO-P_2O_5 sol-gel glasses, the main factor that contributes to the crystallisation of an apatite phase on the surface is the CaO content: *the higher the CaO content the faster the CHA crystallisation*. However, highly mesoporous bioactive glasses show a particular CHA crystallisation kinetic. In these materials, the main factor seems to be the surface area. For materials with lower CaO content but having higher surface area values, the CHA crystallisation is observed at shorter times when soaked in SBF.

This is a very interesting property for material for bone filling and regeneration, because one of the problems of bioactive sol-gel glasses is their "excessive" reactivity due to the initial burst effect of Ca^{2+} release. The intense ionic exchange during the first stages of the bioactive process leads to local pH increase. Depending on the sink conditions of the area (blood perfusion, mainly) the pH increase can be toxic or nontoxic for the surrounding tissues. With these materials, bioactive glasses can be designed with low Ca^{2+} content while maintaining excellent bioactive behaviour.[217]

3.7.8 Biomimetism Evaluation on Silica-Based Bioactive Glasses

Once a protocol (biomimetic solution, dynamic or static test, *etc.*) has been established, the evolution of the bioglass surfaces can be verified by several techniques. In the same way that CaP-derived bioceramics are studied, FTIR spectroscopy is one of the most widely used method to evaluate the biomimetic growth on silica-based glasses. Figure 3.17 indicates the formation of an apatite-like layer on the glass surface after soaking in SBF. Silicate absorption bands at about 1085, 606 and 462 cm^{-1} are observed on the glass spectra before soaking. Phosphate absorption bands at about 1043, 963, 603, 566 and 469 cm^{-1} and carbonate absorption bands at approximately 1490, 1423 and 874 cm^{-1} can be observed in the spectra of materials scraped from the surfaces of soaked glass disks. The increase on the intensity of the carbonate bands is associated with the soaking period in SBF solution. The phosphate and carbonate absorption bands observed on the glass surfaces after soaking are similar to those observed in synthetic carbonate hydroxyapatite.[181] These bands not only confirm the formation of an apatite-like layer, but also indicate that the apatite-like layer material is a carbonate hydroxyapatite similar to biological apatites, in which a coupled substitution of Na^+ by Ca^{2+} and CO_3^{2-} by PO_4^{3-} is observed.[218,219]

The changes in the bioglass surface can also be monitored by X-ray diffraction. Given the amorphous nature of the glass, and its evolution towards an apatite of very low crystallinity, firstly it does not seem a very adequate technique for such a study. However, it is a very useful tool to visualise the transformations on the glass when in contact with SBF, following the evolution

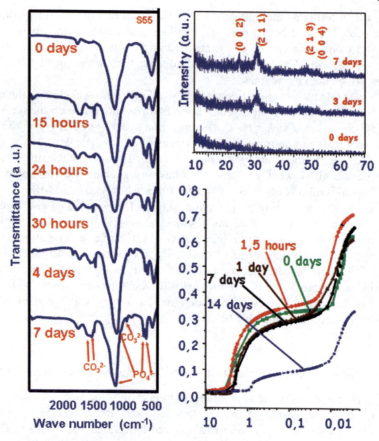

Figure 3.17 Study of the nucleation and growth of an apatite-like layer on the surface of a bioactive sol-gel glass as a function of soaking time in SBF. Left: Fourier transform infrared spectroscopy (FTIR); Right: X-ray diffraction patterns (upper), Hg intrusion porosimetry (lower).

with soaking time. It also allows comparison of the diffraction patterns obtained with those of natural bone; for soaking times equal to or above seven days, clear similarities can be observed (Figure 3.16). As can be observed, the diffraction patterns of bioactive glasses show two diffuse reflections centred at 2θ values of 26° and 32° that correspond to the hydroxyapatite (002) and (211) reflections, respectively. Even after 7 days of soaking in SBF, the XRD patterns correspond to a material with a very low degree of crystallinity.

The biomimetic growth also induces changes in the textural properties of the substrate. In the case of bioactive pieces these changes can be observed at the macroporous level, since the intergranular spaces are filled as the new apatite phase grows. This evolution can be followed by Hg intrusion porosimetry (Figure 3.17) since the volume of Hg intruded is drastically reduced after 2 weeks in SBF.

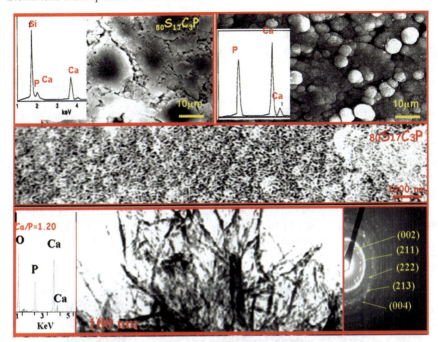

Figure 3.18 Scanning electron micrographs of a bioactive glass before and after soaking in SBF for 7 days (left). HRTEM images of the newly formed biomimetic apatite phase (middle) and electron diffraction (ED) pattern – EDX spectrum of the glass surface after 7 days soaked in SBF (right).

The changes at the bioglass surface can be clearly observed by scanning electron microscopy techniques. Thus, Figure 3.18 shows images of the glass surface after soaking in SBF for one week. These images confirm the formation of a layer constituted of spherical particles, which coats the whole surface of the initial glass. It can be observed that the particles are formed by small crystalline aggregates. The combination of SEM and EDS techniques yields additional information about the nature of this newly formed layer. In fact, the EDS profiles of the glass surface after 1 week of soaking in SBF reveals the presence of P and Ca only, with a Ca/P ratio of approximately 1.25. These results support the growth of a layer with similar composition to that of biological apatites.

In turn, the particles observed by SEM can be further studied by TEM and EDS, analysing their composition by means of an EDS equipment connected to the TEM microscope. Figure 3.18 shows the high magnification image, ED pattern and EDS spectrum of particles at the apatite-like layer grown onto the bioglass surface upon 1 week of soaking in SBF. A small area was selected using the microdiffraction technique. The ED pattern obtained showed the presence of diffuse diffraction rings in which the interplanar spacings agreed with those of an apatite-like structure, indicating that crystalline nuclei were embedded in a glassy matrix. In the corresponding micrograph, the needle-like shape of the aggregated crystals forming the spherical particles may be observed. Taking into account the hydroxyapatite lattice parameters ($a = 9.5\,\text{Å}$ and $c = 6.8\,\text{Å}$), and its symmetry (hexagonal, S.G. P6$_3$/m), most likely its unit cells will be arranged along the c-axis. This would justify a preferred orientation that gives rise to an oriented growth along the c-axis and a needle-like morphology, which agrees with the morphology observed by TEM. On the other hand, the EDS spectrum obtained with a TEM microscope showed that the crystals were composed of Ca, P and O, in agreement with that corresponding to biological apatites.

Another interesting aspect of the apatite-like layer is to ascertain its thickness. The combination of SEM and EDS techniques can be very useful in this question. In Figure 3.19, the cross section of 55S glass (55: SiO$_2$ percentage; S: sol-gel) after 15 h of soaking is shown. The EDS spectra inside the glass and on the layer are also included. As observed, the obtained analysis of the inner region agrees with the nominal glass composition, that is 55% SiO$_2$-41% CaO-4%

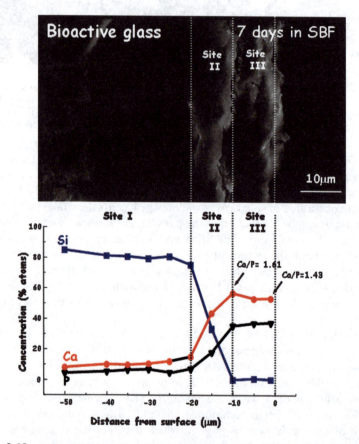

Figure 3.19   SEM micrograph of a cross section of a bioactive glass after being soaked in SBF for 7 days (up). Element distribution obtained by EDX (bottom).

P_2O_5 (in mole%). However, in the EDS spectrum of the layer, a remarkable increase of Ca and P concentrations, together with a significant decrease of Si, was observed. The decrease of Si with increasing Ca and P concentrations indicates the formation of an apatite-like material. On the other hand, the SEM study of the cross section of samples after different soaking times allows monitoring of the evolution of the layer thickness with the soaking time in SBF. Layer thickness grew from 2 μm after 15 h of immersion up to 10 μm after 5 days of assay. It is also observed that there is no difference in layer thickness between 5 and 7 days, which suggests that, at least under *in vitro* conditions, the apatite-like layer does not keep growing indefinitely.

3.8 Biomimetism in Organic-Inorganic Hybrid Materials

3.8.1 An Introduction to Organic-Inorganic Hybrid Materials

Organic-inorganic hybrid materials have the unique feature of combining the properties of traditional materials, such as ceramics and organic polymers, on the nanoscopic scale.[220–228] Nowadays, these materials represent the most direct approach toward the development of an artificial bone, which is to develop materials with similar composition and/or structure in nanodimensional, physical, biochemical and biological response to natural bone.

The synthesis methodology is closely related to the development of the sol-gel science.[229,230] The general behaviour of these organic-inorganic nano-composites is dependent on the nature and relative content of the constitutive inorganic and organic components, although other parameters such as the synthesis conditions also determine the properties of the final materials. The final product must be an intimate "mixture" where at least one of the domains (inorganic or organic) has a dimension ranging from a few angstroms to a few tens of nanometres. In this section, we will review the behaviour of these implants able to mimic some of the functional properties of bone, especially that concerning the production of nanoapatites in contact with physiological fluids.

The main goal when synthesising a silicate-containing hybrid material for any application, including biomedical ones, is to take advantage from both domains to improve the final properties. In Section 3.9, we could see how the silica-based bioactive glasses are able to promote the formation of nanoapatites in contact with physiological fluids. The high bioactivity of silicate-based glasses suggests that the incorporation of silicate as an inorganic component would supply bioactivity to the organic component through the hybrid material synthesis.

The final properties are not only the addition of the properties of the individual components but synergetic effects can be expected according to the high interfacial area. Table 3.8 collects some of the features that each domain can supply to the hybrid.

Table 3.8   Respective properties from the organic and inorganic domains, expected to be combined in hybrid meterials.

Inorganic	Organic
- Hardness, brittleness	- Elasticity, plasticity
- Strength	- Low density
- Thermal stability	- Gas permeability
- High density	- Hydrophobicity
- High refractive index	- Selective complexation
- Mixed valence state (red-ox)	- Chemical reactivity
- Bioactivity	- . . .

Based on the nature of the interactions exchanged by both components, organic-inorganic hybrid materials can be classified as class I and II.[231] Class I hybrid materials show weak interactions between both domains, such as van der Waals, hydrogen bonds and electrostatic interactions. No chemical links (covalent or iono-covalent) are present between the components. In these cases, silica is considered as an inorganic nanofiller incorporated into the organic component. On the contrary, class II organic-inorganic hybrid materials show chemical links between the components and, consequently strong interactions are produced. In this last case, the silicates are considered to be organically modified and they are usually referred to as *ormosils*.

3.8.2   Synthesis of Biomimetic Nanoapatites on Class I Hybrid Materials

The possibility to design class I hybrid materials associating biopolymers with mineral phases relies on the understanding and control of their mutual interaction. An interesting approach is synthesising organic-inorganic hybrids based on bioactive gel glasses (BG) and a biocompatible hydrophilic organic polymer such as poly(vinyl alcohol) (PVAL). The synthesis of BG-PVAL-based hybrid materials aims to obtain a new family of compounds, which exhibits the bioactive behaviour of sol-gel glasses together with the mechanical properties and biodegradability of PVAL. The bioactive glass component can belong to the SiO_2-CaO-P_2O_5 or SiO_2-CaO systems. The presence of these kinds of components not only ensures the implant integration, but also stimulates the new bone formation due to the action of their degradation products (soluble silica, Ca^{2+} cations, *etc.*) on the gene expression of bone-growth factors.

These systems can be synthesised as monoliths, being potentially applicable for the treatment of medium and large bone defects. When the biodegradability and bioactivity of these hybrids were studied after being soaked in SBF, it could be observed that the addition of PVAL helped the synthesis of crack-free monoliths able to develop an apatite-like phase.[232–233] On the contrary, higher amounts of P_2O_5 made the hybrid synthesis difficult and decreased their *in vitro* bioactivity, although it also contributes to the material degradability. Thus, hybrids with very large amounts of both PVAL and P_2O_5 showed such a fast degradation that apatite formation is impeded.

3.8.3 Synthesis of Biomimetic Nanoapatites on Class II Hybrid Materials

The strategy to synthesise class II hybrid materials consists of making intentionally strong bonds (covalent or iono-covalent) between the organic and inorganic components. Organically modified metal alkoxides are hybrid molecular precursors that can be used for this purpose,[234] but the chemistry of hybrid organic-inorganic networks is mainly developed around silicon-containing materials. Currently, the most common way to introduce an organic group into an inorganic silica network is to use organo-alkoxysilane molecular precursors or oligomers of general formula $R'_nSi(OR)_{4-n}$ or $(OR)_{4-n}Si-R''-Si(OR)_{4-n}$ with $n = 1,2,3$. The sol-gel synthesis of siloxane-based hybrid organic-inorganic implants usually involves di- or trifunctional organosilanes cocondensed with metal alkoxides, mainly $Si(OR)_4$ and $Ti(OR)_4$. Finally, the Ca salt incorporation is a common strategy to provide bioactivity at the systems.

3.8.3.1 PMMA-Silica Ormosils

PMMA-silica hybrid composites have been prepared for dental-restorative and bone-replacement applications.[235,236] This hybrid material exhibits growth of a low-crystalline CHA layer on the surface when soaked in SBF, pointing out the bioactive behaviour of this hybrid. Biocompatibility tests have been carried out with these kinds of materials.[237] Mouse calvarial osteoblast cell cultures showed better biological response when seeded on PMMA-SiO_2 hybrid materials than on PMMA in terms of cell attachment, proliferation and differentiation. The enhanced biocompatibility of the PMMA-SiO_2 hybrid was explained by two possible interrelated mechanisms: a) the capability of inducing a calcium phosphate layer formation on the surface of the PMMA-SiO_2 in cell culture media and b) the capability to release silica (as silicic acid), which induces osteoblast early mineralisation.

3.8.3.2 PEG-SiO$_2$ Ormosils

Poly(ethylene glycol)-SiO_2 ormosils have been prepared as an approach to the preparation of biologically active polymer-apatite composites. For this purpose, Yamamoto *et al.*[238] obtained these class II hybrids from triethoxysilyl-terminated poly(oxyethylene) (PEG) and tetraethoxysilane (TEOS) by using the *in-situ* sol-gel process. After being subjected to the biomimetic process for forming the bone-like apatite layer, it was found that a dense apatite layer could be prepared on the hybrid materials, indicating that the formed silanol groups provide the effective sites for the CHA nucleation and growth.

3.8.3.3   *PDMS-CaO-SiO$_2$-TiO$_2$ Ormosils*

One of the more thoroughly studied organic-inorganic hybrid systems for bone and dental repairing is that including poly(dimethylsiloxane) (PDMS) as precursor, together with titanium or silicon alkoxides such as tetra-ethylorthotitanate (TEOT) or TEOS, respectively. These hybrid materials show properties comparable to those of organic rubbers.[221,239,240]

Chen *et al.*[241,242] have extensively worked on the PDMS-modified CaO-SiO$_2$-TiO$_2$ system, obtaining dense and homogeneous monoliths composed of a silica and titania network incorporated with PDMS and the calcium ion ionically bonded to the network. The hybrids show relatively large amounts of calcium in their surfaces and an apatite-like phase is developed within 12 to 24 h in SBF. Together with this fairly high apatite-forming ability, some compositions of PDMS-CaO-SiO$_2$-TiO$_2$ ormosils exhibit high extensibilities and Young's modulus almost equal to that of the human cancellous bone, although all these features also depends on synthesis parameters such as the thermal treatment.[243]

3.8.3.4   *PDMS-CaO-SiO$_2$ Ormosils*

PDMS-CaO-SiO$_2$ ormosils combine in a single material the excellent bioactivity of the inorganic component, CaO-SiO$_2$, and the rubber-like mechanical properties induced by the organic constituent, PDMS. As might be expected, the bioactive behaviour is strongly dependent on the CaO content.[244] The apatite-forming ability of the hybrids appears when the calcium content in the CaO/SiO$_2$ molar ratio falls into the range of 0–0.1. The hybrids with a CaO/SiO$_2$ molar ratio between 0.1–0.2 formed apatite on their surfaces in SBF within 12 h. These ormosils also showed mechanical properties analogous to those of human cancellous bones.

HRTEM of this hybrid material (Figure 3.20) shows the characteristic contrast distribution observed for amorphous materials, suggesting similar structural features to those of glasses.[203] EDS microanalysis results show the incorporation of Ca atoms randomly distributed into the SiO$_2$ cluster network. The nanostructural analysis revealed distances of 0.53 nm between the [SiO$_4^{4-}$] units. Besides, non-bioactive CaO-SiO$_2$-PDMS materials were also synthesised. For this synthesis, the same amounts of reactants and catalyst as for the bioactive one were used, but in this case twice the amount of H$_2$O was used. The corresponding Fourier-filtered HRTEM image showed an average distance of 0.39 nm between [SiO$_4^{4-}$] units. This distance is clearly lower than 0.53 nm measured for the bioactive hybrid, suggesting that Ca is not incorporated in the nonbioactive material. Since both hybrids exhibit different kinetics of bioactive response, this behaviour can be explained in terms of both nanostructure and chemical composition.

3.8.4   Bioactive Star Gels

In 1995 DuPont Corp. developed the *star-gel* materials.[245–247] *Star gels* are a type of organic-inorganic hybrids that present a singular structure of an organic

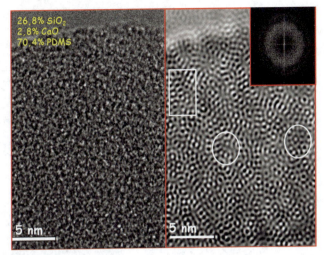

Figure 3.20   Electron microscopy study of a PDMS-SiO$_2$-CaO ormosil: (Left) Original HRTEM image of the amorphous matrix (centre) filtered HRTEM image and (right) Fourier transform pattern. Distances up to 0.53 nm for (SiO$_4$)$^{4-}$ can be observed in the filtered image, indicating the Ca^{2+} presence between tetrahedra.

core surrounded by flexible arms, which are terminated in alkoxysilane groups (see Figure 3.21). At the macroscopic level, *star gels* exhibit behaviour between conventional glasses and highly crosslinked rubbers in terms of mechanical properties. Currently, *star gels* are still one of the most interesting subjects in the field of hybrid materials due to their mechanical properties.[248]

Very recently, the synthesis of *bioactive star-gels* (BSG), *i.e.* star gels capable of integrating with bone tissue, has been developed. As for many other class II hybrid materials, bioactive *star gels* (BSG) are obtained by hydrolysis and condensation of alkoxysilane-containing precursors. In fact, star gels are formulated as single-component molecular precursors with flexibility built in at the molecular level. The starting materials comprise an organic core with multiple flexible arms that terminate in network-forming trialkoxysilane groups. The core can be a single silicon atom, linear disiloxane segment, or ring system, as can be seen in Figure 3.20.

The development of bioactive star gels is still in process. Only the precursors marked as A and B in Figure 3.20 have been used so far, for the design of bioactive implants.[249] The basis of the star-gel's bioactivity consists of incorporating Ca^{2+} cations into the inorganic component of the hybrid structure, thus exhibiting similar properties to conventional SiO$_2$-CaO sol-gel glasses but having the flexibility supplied by the organic chains.

Not all the Ca^{2+}-containing star gels are bioactive. The relative amount of network formers (alkoxysilanes) and network modifiers (Ca^{2+} cations) determine the bioactive behaviour of star gels. More specifically, the Si/Ca ratio provides a good approximation to predict whether a star gel will be bioactive or

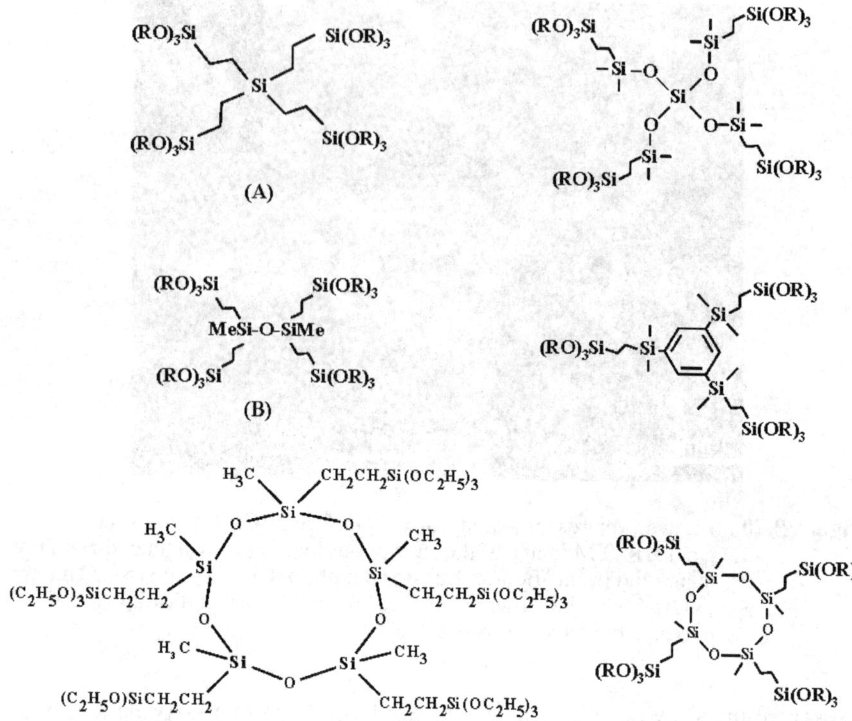

Figure 3.21   Star-gel precursors.

not. All those compositions with Si/Ca ratios higher than 9 are not bioactive, due to the high stability of these star gels at physiological pH. The chemical composition and structure of the precursors must be known, since the number of Si atoms per unit formula must be determined. All the Si atoms must be taken into account and not only those with hydrolisable groups, such as –Si–O–R. In this way, precursors A and B of Figure 3.21 contribute with their 5 and 9 Si atoms, respectively, to the Si/Ca ratio. Figure 3.22 is an example of the surface evolution for a star gel obtained from precursor A and with a Si/Ca ratio of 5. This figure shows the scanning electron micrographs for sample SGA-Ca before and after soaking in SBF for 7 and 17 days. Before soaking, the micrograph shows a smooth surface characteristic of a nonporous and homogeneous gel. Besides, EDX spectroscopy confirms the presence of Si and Ca as the only components of the inorganic phase. After 7 days in SFB, a new phase partially covers the *star-gel* surface. This phase is formed by rounded submicrometre particles composed of Ca and P, as EDX spectroscopy indicates. After 17 days in SBF, the monolith surface is fully covered by a layer constituted of spherical particles, which are formed by numerous needle-shaped crystallites (character- istic of the apatite phase growth over bioactive materials surface). At this point, the EDX spectrum indicates that the surface is fully covered by a

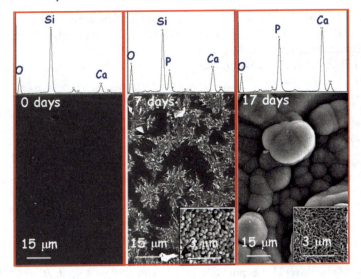

Figure 3.22 SEM micrographs and EDX spectra of the SGA-Ca surface before and after being soaked in SBF.

calcium phosphate with a Ca/P ratio of 1.6, *i.e.* that corresponding to a calcium-deficient apatite.

Bioactive star-gels can be excellent candidates for bone-tissue regeneration since they fulfil the following features: a) Easily obtained as monoliths of different shapes in order to fit to any kind of medium or large bone defect; b) structurally homogeneous to predict their biological and mechanical response when implanted; c) able to develop an apatite-like phase in contact with physiological fluids, *i.e.* must be bioactive and d) mechanical properties significantly better than those exhibited by conventional bioactive glasses.

References

1. R. L. Reis, *Curr. Opin. Solid State Mater. Sci.*, 2003, **7**, 263.
2. M. Vallet-Regí and D. Arcos, *Nanostructured Hybrid Materials for Bone Implants Fabrication*, In: *Bioinorganic Hybrid Nanomaterials*, E. Ruiz-Hitzky, K. Ariga and Y. M. Lvov eds., Wiley-VCH Verlag GmbH & Co. KGaA, Weinheim, 2007.
3. S. V. Dorozhkin and M. Epple, *Angew. Chem. Int. Ed.*, 2002, **41**, 3130.
4. T. Kokubo, H. Kushitani, S. Sakka, T. Kitsugi and T. J. Yamamuro, *Biomed. Mater. Res.*, 1990, **24**, 721.
5. W. E. Brown, N. Eidelman and B. Tomazic, *Adv. Dent Res.*, 1987, **1**, 306.
6. W. E. Brown and M. U. Nylen, *J. Dent. Res.*, 1964, **43**, 751.
7. L. L. Hench, R. J. Splinter, T. K. Greenly and W. C. Allen, *J. Biomed. Mater. Res.*, 1971, **2**, 117.

8. T. Kokubo, K. Hata, T. Nakamura and T. Yamamuro, in *Bioceramics*, W. Bonfield, G. W. Hastings and K. E. Tunner eds., Butterworth-Heinemann, UK, vol 4., p 113.

9. M. Tanahashi, T. Kokubo and T. Matsuda, *J. Biomed. Mater. Res.*, 1996, **31**, 243.

10. H. Ohgushi and A. I. Caplan, *J. Biomed. Mater. Res.*, 1999, **48**, 913.

11. S. Leeuwenburgh, P. Layrolle, F. Barrere, J. de Bruijn, J. Schoonman, C. A. van Blitterswijk and K. de Groot, *J. Biomed. Mater. Res.*, 2001, **56**, 208.

12. L. L. Hench, *J. Am. Ceram. Soc.*, 1991, **74**, 1487.

13. S. Fujibayashi, M. Neo, J. M. Kim, T. Kokubo and T. Nakamura, *Biomaterials*, 2003, **24**, 1349.

14. T. Kokubo, H. Kushitani, C. Ohtsuki, S. Sakka and T. Yamamuro, *J. Mater. Sci.: Mater. Med.*, 1992, **3**, 79.

15. C. Du, P. Klasens, R. E. Haan, J. Bezemer, F. Z. Cui, K. de Groot and P. Layrolle, *J. Biomed Mater. Res.*, 2002, **59**, 535.

16. A. M. Radder, H. Leenders and C. A. van Blitterswijk, *J. Biomed. Mater. Res.*, 1994, **28**, 141.

17. A. M. Radder, J. E. Davies, J. Leeners and C. A. van Blitterswijk, *J. Biomed. Mater. Res.*, 1994, **28**, 269.

18. M. Okumura, C. A. Blitterswijk, H. K. Koerten, D. Bakker, K. De Groot and H. Ohgushi, *Advances in Biomaterials*, P. L. Doherty, R. L. Williams, D. F. Williams and A. J. C. Lee eds., Elsevier, Amsterdam, 1992, Vol. 10, pp. 343–347.

19. C. A. van Blitterswijk, J. van den Brink, H. Leenders and D. Bakker, *Cells Mater.*, 1993, **5**, 55.

20. A. M. Radder and C. A. van Blitterswijk, *J. Mater. Sci.: Mater. Med.*, 1994, **5**, 320.

21. A. M. Radder, J. E. Davies, R. N. S. Sodhi, S. A. T. van der Meer, J. G. C. Wolke and C. A. van Blitterswijk, *Cells Mater.*, 1995, **5**, 320.

22. G. J. Meijer, A. van Dooren, M. L. Gaillard, R. Dalmeijer, C. De Putter and C. A. van Blitterswijk, *Int. J. Oral Maxillofac. Surg.*, 1996, **25**, 210.

23. C. Du, G. J. Meijer, C. Van de Valk, R. E. Haan, J. M. Bezemer, S. C. Hesseling, F. Z. Cui, K. De Groot and P. L. Layrolle, *Biomaterials*, 2002, **23**, 4649.

24. K. de Groot, R. G. T. Geesink, C. P. A. T. Klein and P. Serekian, *J. Biomed. Mater. Res.*, 1987, **21**, 1375.

25. W. L. Jaffe and D. F. Scott, *J Bone Joint Surg.*, 1996, **78A**, 1918.

26. P. Li, *J. Biomed. Mater. Res.*, 2003, **66A**, 79.

27. Y. F. Chou, I. Wulur, J. C. Y. Duna and B. J. Wu, *Handbook of Nanostructured Biomaterials and their Applications in Nanobiotechnology*, H. S. Nalwa ed., American Scientific Publishers, Stevenson Ranch, 2005, Vol. 2., pp 197–222.

28. T. Kokubo, H. M. Kim and M. Kawashita, *Biomaterials*, 2003, **245**, 2161.

29. H. M. Kim, *Curr. Opin. Solid State Mater. Sci.*, 2003, **7**, 289.

30. S. V. Dorozhkin, *J. Mater. Sci.*, 2007, **42**, 1061.

31. S. Radin and P. Ducheyne, *J. Biomed. Mater. Res.*, 1996, **30**, 273.
32. M. S. A. Johnsson, E. Paschalis and G. H. Nancollas, In: *Bone-biomaterial Interface*, J. E. Davies ed., Toronto, University of Toronto Press, 1991. p. 62–75.
33. R. I. Martin and P. W. Brown, *Mater. Med.*, 1994, **5**, 96.
34. T. Kokubo T. H. Kushitani, S. Sakka, T. Kitsugi, S. Kotani, K. Oura and T. Yamamuro, *Apatite formation on bioactive ceramics in body environment*, In: *Bioceramics*, H. Oonishi, H. Aoki and K. Sawai eds., Tokyo: Ishiyaku Euro America, Inc., 1989, Vol. 1., pp. 157–162.
35. H. M. Kim, F. Miyaji, T. Kokubo and T. Nakamura, *J. Ceram. Soc. Jpn.*, 1997, **105**, 111.
36. T. Miyazaki, H. M. Kim, F. Miyaji, T. Kokubo, H. Kato and T. Nakamura, *J. Biomed. Mater. Res.*, 2000, **50**, 35.
37. Y. Abe, T. Kokubo and T. Yamamuro, *J. Mater. Sci.: Mater. Med.*, 1990, **1**, 233.
38. M. Tanahashi, T. Yao, T. Kokubo, M. Minoda, T. Miyamoto, T. Nakamura and T. Yamamuro, *J. Am. Ceram. Soc.*, 1994, **77**, 2805.
39. A. Oyane, M. Minoda, T. Miyamoto, K. Nakanishi, M. Kawashita, T. Kokubo, T. Nakamura, Apatite formation on ethylene–vinyl alcohol copolymer modified with silane coupling agent and calcium silicate, In: *Bioceramics*, S. Giannini and A. Moroni eds., Vol. 13. Trans Tech Publications, Zurich, 2000. p 713–716.
40. H. M. Kim, K. Kishimoto, F. Miyaji, T. Kokubo, T. Yao, Y. Suetsugu, J. Tanaka and T. Nakamura, *J. Biomed. Mater. Res.*, 1999, **46**, 228.
41. H. M. Kim, K. Kishimoto, F. Miyaji, T. Kokubo, T. Yao, Y. Suetsugu, J. Tanaka and T. Nakamura, *J. Mater. Sci.: Mater. Med.*, 2000, **11**, 421.
42. W. F. Newman and M. W. Newman, *The Chemical Dynamics of Bone Mineral*, The University of Chicago Press, Chicago, 1967, p.18.
43. A. Oyane, H. M. Kim, T. Furuya, T. Kokubo and T. Miyazaki, *J. Biomed. Mater. Res.*, 2003, **65A**, 188.
44. H. Takadama, M. Hashimoto, M. Mizuno and T. Kokubo, *Phos. Res. Bull.*, 2004, **17**, 119.
45. T. Hanaba, K. Asami and K. Asaoka, *J. Biomed. Mater. Res.*, 1998, **40**, 530.
46. C. A. Homsy, *J. Biomed. Mater. Res.*, 1970, **4**, 341.
47. K. Hyakuna, T. Yamamuro, Y. Kotoura, M. Oka, T. Nakamura, T. Kitsugi, T. Kokubo and H. Kushitani, *J. Biomed. Mater. Res.*, 1990, **24**, 471.
48. A. C. Lewis, M. R. Kilburn, I. Papageorgiou, G. C. Allen and C. P. Case, *J. Biomed. Mater. Res*, 2005, **73A**, 456.
49. Y. Gao, W. Weng, K. Cheng, P. Du, G. Shen, G. Han, B. Guan and W. Yan, *J. Biomed. Mater. Res.*, 2006, **79A**, 193.
50. F. Miyaji, H. M. Kim, S. Handa, T. Kokubo and T. Nakamura, *Biomaterials*, 1999, **20**, 913.
51. F. Barrere, C. A. van Blitterswijk, K. de Groot and P. Layrolle, *Biomaterials*, 2002, **23**, 2211.
52. A. C. Tas and S. B. Bhaduri, *J. Mater. Res.*, 2004, **19**, 2742.

53. F. Barrere, C. A. van Blitterswijk, K. de Groot and P. Layrolle, *Biomaterials*, 2002, **23**, 1921.
54. F. Barrere, P. Layrolle, C. A. Van Blitterswijk and K. De Groot, *J. Mater. Sci.: Mater. Med.*, 2001, **12**, 529.
55. Y. F. Chou, W. Huang, J. C. Y. Dunn, T. Miller and B. M. Wu, *Biomaterials*, 2005, **26**, 285.
56. L. D. Warren, A. E. Clark and L. L. Hench, *J. Biomed. Mater. Res. Appl. Biomat.*, 1989, **23**, 201.
57. J. H. Hanks and R. E. Wallace, *Proc. Soc. Exp. Biol. Med.*, 1949, **71**, 196.
58. Y. Shibata, H. Takashima, H. Yamamoto and T. Miyazaki, *Int. J. Oral Maxillofac. Implants*, 2004, **19**, 177.
59. P. A. P. Marques, A. P. Serro, B. J. Saramago, A. C. Fernandes, M. C. Magalhaes and R. N. Correia, *Biomaterials*, 2003, **24**, 451.
60. M. Vallet-Regí, A. J. Salinas and D. Arcos, *J. Mater. Sci.: Mater. Med.*, 2006, **17**, 1011.
61. I. Izquierdo-Barba, A. J. Salinas and M. Vallet-Regí, *J. Biomed. Mater. Res.*, 2000, **51**, 191.
62. J. Hlavac, D. Rohanova and A. Helebrant, *Ceram. Silicate*, 1994, **38**, 119.
63. S. Falaize, S. Radin and P. Ducheyne, *J. Am. Ceram. Soc.*, 1999, **82**, 969.
64. A. J. Salinas, M. Vallet-Regí and I. Izquierdo-Barba, *J. Sol-Gel. Sci. Tech.*, 2001, **21**, 13.
65. G. H. Nancollas and W. Wu, *J. Cryst. Growth*, 2000, **211**, 137.
66. P. Koutsoukos, Z. Amjad, M. B. Tomson and G. H. Nancollas, *J. Am. Chem. Soc*, 1980, **102**, 1553.
67. M. B. Tomson and G. H. Nancollas, *Science*, 1978, **200**, 1059.
68. R. Kniep and S. Bush, *Angew. Chem. Int. Ed.*, 1996, **35**, 2624.
69. S. Busch, H. Dolhaine, A. Dúchense, S. Heinz, O. Hochrein, F. Laeri, O. Podebrad, U. Vietze, T. Weiland and R. Kniep, *Eur. J. Inorg. Chem.*, 1999, 1643.
70. S. Busch, U. Schwarz and R. Kniep, *Chem. Mater.*, 2001, **13**, 3260.
71. H. Tlatlik, P. Simon, A. Kawska, D. Zahn and R. Kniep, *Angew. Chem. Int. Ed.*, 2006, **45**, 1905.
72. P. Simon, D. Zahn, H. Lichte and R. Kniep, *Angew. Chem. Int. Ed.*, 2006, **45**, 1911.
73. D. Zaffe, *Micron*, 2005, **36**, 583.
74. E. L. Burger and V. Patel, *Orthopedics*, 2007, **30**, 939.
75. W. Suchanek and M. Yoshimura, *J. Mater. Res*, 1998, **13**, 94.
76. M. Neo, T. Nakamura, T. Yamamuro, C. Ohtsuki and T. Kokubo, In: *Bone-bonding Biomaterials*, P. Ducheyne, T. Kokubo and C. A. van Blitterswijk eds., Reed Healthcare Communications, Leiderdorp, Netherlands, 1993, p.111–120.
77. P. Ducheyne, J. Beight, J. Cuckler, B. Evans and S. Radin, *Biomaterials*, 1990, **11**, 531.
78. P. Ducheyne and J. M. Cuckler, *Clin. Orthop. Rel. Res.*, 1992, **276**, 102.
79. J. D. de Bruijn, Y. P. Novell and C. A. van Blitterswijk, *Biomaterials*, 1994, **15**, 543.

80. S. H. Maxian, J. P. Zawadski and M. G. Duna, *J. Biomed. Mater. Res.*, 1993, **27**, 111.
81. M. Jarcho, J. F. Kay, K. I. Gumaer, R. N. Doremus and H. P. Drobeck, *J. Bioeng.*, 1977, **1**, 79.
82. B. M. Tracy and R. H. Doremus, *J. Biomed. Mater. Res.*, 1984, **18**, 719.
83. G. Daculsi, R. Z. LeGeros, E. Nery, K. Lynch and B. Kerebel, *J. Biomed. Mater. Res.*, 1988, **23**, 257.
84. S. Langstaff, M. Sayer, T. Smith, S. Pugh, S. Hesp and W. Thompson, *Biomaterials*, 2001, **22**, 135.
85. K. Kurashina, H. Kurita, M. Hirano, A. Kotani, C. P. Klein and D. de Groot, *Biomaterials*, 1997, **18**, 539.
86. S. Takagi, L. C. Chow and K. Ishikawa, *Biomaterials*, 1998, **19**, 1593.
87. P. Ducheyne and Q. Qiu, *Biomaterials*, 1999, **20**, 2287.
88. R. Z. Le Geros, J. R. Parsons, G. Daculsi, F. Driessens, D. Lee, S. T. Liu, S. Metsger, D. Peterson, M. Walker, in *Bioceramics: Material Characteristics Versus In vivo Behavior*, P. Ducheine and J. Lemons, eds., *N.Y. Acad. Sci.*, 1988, **523**, 268–271.
89. T. Fujui and M. Ogino, *J. Biomed. Mater. Res*, 1984, **18**, 845.
90. L. L. Hench, "Bioactive Ceramics," in *Bioceramics: Material Characteristics Versus In vivo Behaviour*, P. Ducheyne and J. Lemons eds., *N.Y. Acad. Sci.*, 1988, **54**, 523.
91. P. Ducheyne, S. Radin and L. King, *J. Biomed. Mater. Res.*, 1993, **27**, 25.
92. A. S. Posner, *Clin Orthop.*, 1985, **200**, 87.
93. W. van Raemdonck, P. Ducheyne and P. de Meester, in *Metal and Ceramic Biomaterials*, P. Ducheyne and W. Hasting eds., CRC Press, Boca Raton, 1984, p. 149.
94. M. Jarcho, *Clin. Orthop.*, 1981, **157**, 259.
95. J. C. Elliott, *Structure and Chemistry of the Apatites and other Calcium Orthophosphates*. Elsevier, Amsterdam, 1994.
96. R. Z. LeGeros, *Calcium Phosphates in Oral Biology and Medicine. Karger*, Basel, 1991.
97. Y. Leng, J. Chen and S. Qu, *Biomaterials*, 2003, **24**, 2125.
98. S. R. Radin and P. Ducheyne, *J. Biomed. Mater. Res.*, 1993, **27**, 35.
99. R. H. Doremus, *J. Mater. Sci.*, 1992, **27**, 285.
100. K. A. Gross and C. C. Berndt, *J. Biomed. Mater. Res.*, 1998, **39**, 580.
101. T. Kobayashi, S. Shingaki, T. Nakajima and K. Hanada, *J. Long-Term Effects Med. Impl.*, 1993, **3**, 283.
102. W. Bonfield, M. D. Grynpas, A. E. Tuly, J. Bowman and J. Abram, *Biomaterials*, 1981, **2**, 185.
103. A. Sari, R. Yavuzer, S. Ayhan, S. Tuncer, O. Latifoglu, K. Atabay and M. C. Celebi, *J. Craniofac. Surg.*, 2003, **14**, 919.
104. M. C. Kruyt, W. J. A. Dhert, C. Oner, C. A. van Blitterswijk, A. J. Verbout and J. D. de Bruijn, *J. Biomed. Mater. Res.*, 2004, **69B**(2), 113.
105. S. F. Hulbert, L. L. Hench, D. Forbers and L. S. Bowman, *Ceram. Int.*, 1982, **8**, 121.

106. C. Ergun, T. J. Webster, R. Bizios and R. H. Doremus, *J. Biomed. Mater. Res.*, 2002, **59**, 305.

107. R. A. Young and P. E. Mackie, *Mater. Res. Bull*, 1980, **15**, 17.

108. R. M. Wilson, J. C. Elliott and S. E. P. Dowker, *Am. Miner.*, 1999, **84**, 1406.

109. E. A. P. De Maeyer, R. M. H. Verbeeck and D. E. Naessens, *Inorg. Chem.*, 1993, **32**, 5709.

110. R. M. H. Verbeeck, E. A. P. De Maeyer and F. C. M. Driessens, *Inorg. Chem.*, 1995, **34**, 2084.

111. E. M. Carlisle, *Science*, 1970, **167**, 179.

112. E. M. Carlisle, *Calc. Tissue Int.*, 1981, **33**, 27.

113. L. L. Hench and G. P. LaTorre, in *Bioceramics 5*, T. Yamamuro, T. Kokubo and T. Nakamura eds., Kobunshi Kankokai, Inc., Kyoto, 1993, pp. 67–74.

114. C. Ohtsuki, T. Kokubo and T. Yamamuro, *J. Non-Cryst. Solids*, 1992, **143**, 84.

115. D. Arcos, D. C. Greenspan and M. Vallet-Regí, *Chem. Mater.*, 2002, **14**, 1515.

116. D. Arcos, D. C. Greenspan and M. Vallet-Regí, *J. Biomed. Mater. Res.*, 2003, **65A**, 344.

117. I. R. Gibson, J. Huang, S. M. Best and W. Bonfield, in *Bioceramics 12*, H. Ohgushi, G. W. Hastings and T. Yoshikawa eds., World Scientific, Singapore, 1999, pp. 191–194.

118. K. A. Hing, S. Saeed, B. Annaz, T. Buckland and P. A. Revell, *Transactions 7th World Biomaterials Congress*, Australian Society for Biomaterials, Brunswick Lower, Vic., 2004, p. 108.

119. I. R. Gibson, S. M. Best and W. Bonfield, *J. Biomed. Mater. Res.*, 1999, **44**, 422.

120. S. M. Best, W. Bonfield, I. R. Gibson, L. J. Jha and J. D. Santos, International Patent Appl. No. PCT/GB97/02325, 1996.

121. I. R. Gibson, S. M. Best and W. Bonfield, *J. Am. Ceram. Soc.*, 2002, **85**, 2771.

122. N. Rashid, I. Harding and K. A. Hing, *Transactions 7th World Biomaterials Congress*, Australian Society for Biomaterials, Brunswick Lower, Vic., 2004, p. 106.

123. S. R. Kim, J. H. Lee, Y. T. Kim, D. H. Riu, S. J. Jung, Y. J. Lee, S. C. Chung and Y. H. Kim, *Biomaterials*, 2003, **24**, 1389.

124. P. A. Marques, M. C. F. Magalhaes, R. N. Correia and M. Vallet-Regí, *Key Eng. Mater.*, 2001, **192–195**, 247.

125. A. J. Ruys, *J. Aust. Ceram. Soc.*, 1993, **29**, 71.

126. S. R. Kim, D. H. Riu, Y. J. Lee and Y. H. Kim, *Key Eng. Mater.*, 2002, **218–220**, 85.

127. D. Arcos, J. Rodriguez-Carvajal and M. Vallet-Regí, *Chem. Mater.*, 2004, **16**, 2300.

128. M. Vallet-Regí and D. Arcos, *J. Mater. Chem.*, 2005, **15**, 1509.

129. D. Arcos, J. Rodriguez-Carvajal and M. Vallet-Regí, *Solid State Sci.*, 2004, **6**, 987.
130. D. Arcos, J. Rodriguez-Carvajal and M. Vallet-Regí, *Physica B*, 2004, **350**, e607.
131. D. Arcos, J. Rodriguez-Carvajal and M. Vallet-Regí, *Chem. Mater.*, 2005, **17**, 57.
132. J. R. Hupp and S. J. McKenna, *J. Oral Maxillofac. Surg.*, 1988, **46**, 533.
133. M. El Deeb and M. Roszkowski, *J. Oral Maxillofac. Surg.*, 1988, **46**, 33.
134. B. V. Rejda, J. G. J Peelen and K. de Groot, *J. Bioeng.*, 1977, **1**, 93.
135. M. Vallet-Regí, *J. Chem. Soc. Dalton Trans.*, 2001, 97.
136. R. Ellinger, E. B. Nery and K. L. Lynch, *Int. J. Periodont. Tertor. Dent.*, 1986, **3**, 23.
137. A. Takeishi, H. Hayashi, H. Kamatsubara, A. Yokoyama, M. Kohri, T. Kawasaki, K. Micki and T. Kohgo, *J. Dent. Res.*, 1989, **68**, 680.
138. G. Daculsi, N. Passuti, S. Martin, C. Deudon, R. Z. LeGeros and S. Rather, *J. Biomed. Mater. Res.*, 1990, **24**, 379.
139. C. Schopper, F. Ziya-Ghazvini, W. Goriwoda, D. Moser, F. Wanschitz, E. Spassova, G. Lagogiannis, A. Auterith and R. Ewers, *J .Biomed Mater. Res. Appl. Biomater.*, 2005, **74B**, 458.
140. M. Trecant, J. Delecrin, J. Royer, E. Goyenvalle and G. Daculsi, *Clin. Mater.*, 1994, **18**, 233.
141. A. Sendemir-Urkmez and R. D. Jamison, *J. Biomed. Mater. Res.*, 2007, **81A**, 624.
142. C. R. Yang, Y. J. Wang, X. F. Chen and N. R. Zhao, *Mater. Lett.*, 2005, **59**, 3635.
143. S. Sánchez-Salcedo, I. Izquierdo-Barba, D. Arcos and M. Vallet-Regí, *Tissue Eng.*, 2006, **12**, 279.
144. M. I. Alam, I. Asahina, K. Ohmmaiuda and S. Enomoto, *J. Biomed. Mater. Res.*, 2000, **54**, 129.
145. O. Gauthier, J. Guicheux, G. R. Grimandi, A. Faivre-Cahuvet and G. Daculsi, *J. Biomed. Mater. Res.*, 1998, **40**, 606.
146. O. Bermúdez, M. G. Boltong, F. C. M. Driessens and J. A. Planell, *J. Mater. Sci.: Mater. Med.*, 1994, **5**, 160.
147. H. Yamamoto, S. Niwa, M. Hori, T. Hattori, K. Sawai, S. Aoki, M. Hirano and H. Takeuchi, *Biomaterials*, 1998, **19**, 1587.
148. M. Sayer, A. Stratilatov, J. Reid, L. Calderin, M. Stott, X. Yin, M. McKenzie, J. N. Smith, J. A. Hendry and S. D. Langstaff, *Biomaterials*, 2003, **24**, 369.
149. S. Langstaff, M. Sayer, T. Smith, S. Pugh, S. Hesp and W. Thompsom, *Biomaterials*, 1999, **20**, 1727.
150. A. Pietak and M. Sayer, *Biomaterials*, 2005, **24**, 3819.
151. A. Takeishi, H. Hayashi, H. Kamatsubara, A. Yokoyama, M. Kohri, T. Kawasaki, K. Miki and T. Kohgo, *J. Dent. Res.*, 1989, **68**, 680.
152. R. Z. LeGeros, G. Daculsi, E. Nery, K. Lynch and B. Kerebel, *Transactions of the Third World Biomaterials Congress*, 1988, **2B**, 1-35.

153. M. Kohri, K. Miki, D. E. Waite, H. Nakajima and T. Okabe, *Biomaterials*, 1993, **14**, 299.
154. L. L. Hench, *J. Mater. Sci.: Mater. Med.*, 2006, **17**, 967.
155. L. L. Hench, A. E. Clark and H. F. Schaake, *Int. J. Non-Cryst. Sol.*, 1972, **8–10**, 837.
156. L. L. Hench and A. Paschall, *J. Biomed. Mater. Res. Symp.*, 1973, **4**, 25.
157. L. L. Hench, *Curr. Opin. Solid State Mater. Sci.*, 1997, **2**, 604.
158. M. Ogino, F. Ohuchi and L. L. Hench, *J. Biomed. Mater. Res.*, 1980, **14**, 55.
159. U. Gross, R. Kinne, H. J. Schmitz and V. Strunz, In *CRC Critical Reviews in Biocompatibility*, D. L. Williams, ed., CRC Press, Boca Raton, Florida, Vol. 4 (Issue 2), 1988, 155.
160. L. L. Hench, In *Bioceramics: Materials Characteristics Versus In vivo Behaviour*, Vol 523, J. P. Ducheyne and J. Lemmons, eds., Annuals of the New York Academy of Sciences, 1988, 54.
161. R. Hill, *J. Mater. Sci. Lett.*, 1996, **15**, 1122.
162. K. E. Wallace, R. G. Hill, J. T. Pembroke, C. J. Brown and P. V. Hatton, *J. Mater. Sci.: Mater. Med.*, 1999, **10**, 697.
163. O. H. Anderson, K. H. Karlsson and K. Kangasmiemi, *J. Non-Cryst. Solids*, 1990, **119**, 290.
164. D. C. Greenspan and L. L. Hench, *J. Biomed. Mater. Res.*, 1976, **10**, 503.
165. Y. Ebisawa, F. Miyaji, T. Kokubo, K. Ohura and T. Nakamura, *Biomaterials*, 1997, **18**, 1277.
166. M. Brink, *J. Biomed. Mater. Res.*, 1997, **36**, 109.
167. M. Brink, T. Turunen, R.-P. Happonen and A. Yli-Urpo, *J. Biomed. Mater. Res.*, 1997, **37**, 114.
168. R. Li, A. E. Clark and L. L. Hench, *J. Appl. Biomater.*, 1991, **2**, 231.
169. M. Catauro, G. Laudisio, A. Costantini, R. Fresa and F. Branda, *J. Sol-Gel Sci. Technol.*, 1997, **10**, 231.
170. I. Izquierdo-Barba, A. J. Salinas and M. Vallet-Regí, *J. Biomed.Mater. Res.*, 1999, **47**, 243.
171. A. Martínez, I. Izquierdo-Barba and M. Vallet-Regí, *Chem. Mater.*, 2000, **12**, 3080.
172. M. M. Pereira, A. E. Clark and L. L. Hench, *J. Am. Ceram. Soc.*, 1995, **78**, 2463.
173. M. M. Pereira and L. L. Hench, *J. Sol-Gel Sci.*, 1996, **7**, 59.
174. T. Peltola, M. Jokinen, H. Rahiala, E. Levanen, J. B. Rosenholm, I. Kangasniemi and A. Yli-Urpo, *J. Biomed. Mater. Res.*, 1999, **44**, 12.
175. M. Vallet-Regí, D. Arcos and J. Pérez-Pariente, *J. Biomed. Mater. Res.*, 2000, **51**, 23.
176. M. Vallet-Regí and A. Rámila, *Chem. Mater.*, 2000, **12**, 961.
177. D. C. Greenspan, J. P. Zhong and G. P. LaTorre, In *Bioceramics 7, Turku*, O. H. Anderson and A. Yli-Urpo, eds., Butterworth-Heinemann Ltd., Oxford, 1994, p. 55.
178. M. Jokinen, H. Rahiala, J. B. Rosenholm, T. Peltola and I. Kangasniemi, *J. Sol-Gel Sci. Technol.*, 1998, **12**, 159.

179. D. Arcos, C. V. Ragel and M. Vallet-Regí, *Biomaterials*, 2001, **22**, 701.
180. M. Laczka, K. Cholewa and A. Laczka-Osyczka, *J. Alloys Compd.*, 1997, **248**, 42.
181. M. Vallet-Regí, A. M. Romero, C. V. Ragel and R. Z. LeGeros, *J. Biomed. Mater. Res.*, 1999, **44**, 416.
182. M. M. Pereira, A. E. Clark and L. L. Hench, *J. Biomed. Mater. Res.*, 1994, **28**, 693.
183. J. Pérez-Pariente, F. Balas, J. Román, A. J. Salinas and M. Vallet-Regí, *J. Biomed. Mater. Res.*, 1999, **47**, 170.
184. K. Ohura, T. Nakamura, T. Kokubo, Y. Ebisawa, Y. Kotoura and M. Oka, *J. Biomed. Mater. Res.*, 1991, **25**, 357.
185. Y. Ebisawa, T. Kokubo, K. Ohura and T. Yamamuro, *J. Mater. Sci.: Mater. Med.*, 1990, **1**, 239.
186. L. L. Hench and O. Andersson, In *Bioactive Glasses. An Introduction to Bioceramics*, L. L. Hench and J. Wilson, eds., World Scientific Publishing, Singapore, 1993, p. 41.
187. C. Ohtsuki, T. Kokubo, K. Takatsuka and T. Yamamuro, *Nippon Seramikkusu Kyokai Gakujutsu Ronbunshi*, 1991, **99**, 1.
188. L. L. Hench, in: *Ceramics: Towards the 21st Century*, N. Soga and S. Kato eds., (Ceramic Society of Japan, Tokyo, 1991) p. 519.
189. Ö. H. Andersson and K. H. Karlsson, *J. Non-Cryst. Solids*, 1991, **129**, 145.
190. W. D. Kingery, H. K. Bowen and D. R. Bowen, in: *Introduction to Ceramics*, 2nd edn., Wiley, New York, 1960, p. 328.
191. M. Vallet-Regí, C. V. Ragel and A. J. Salinas, *Eur. J. Inorg. Chem.*, 2003, 1029.
192. D. C. Greenspan, J. P. Zhong and G. P. LaTorre, *Bioceramics*, 1995, **8**, 477.
193. D. Arcos, J. Peña and M. Vallet-Regí, *Key Eng. Mater.*, 2004, **254–256**, 27.
194. F. G. Araujo, G. P. Latorre and L. L. Hench, *J. Non-Cryst. Solids*, 1995, **185**, 41.
195. R. Li, A. E. Clark and L. L. Hench, In *Chemical Processing of Advanced Materials*, L. L. Hench and J. K. West, eds., John Wiley and Sons, New York, 1992, p. 627.
196. F. Balas, D. Arcos, J. Pérez-Pariente and M. Vallet-Regí, *J. Mater. Res.*, 2001, **16**, 1345.
197. M. M. Pereira and L. L. Hench, *J. Sol-Gel Sci. Technol.*, 1996, **7**, 231.
198. P. Li, C. Ohtuki, T. Kokubo, K. Nakanishi, N. Soja, T. Nakamura and T. Yamamuro, *J. Am. Ceram. Soc.*, 1992, **75**, 2094.
199. K. H. Karlsson, K. Froberg and T. Ringbom, *J. Non-Cryst. Solids*, 1989, **112**, 69.
200. P. E. Wang and T. K. Chaki, *J. Mater. Sci.: Mater. Med.*, 1995, **6**, 94.
201. M. Vallet-Regí, I. Izquierdo-Barba and A. J. Salinas, *J. Biomed. Mater. Res.*, 1999, **46**, 560.

202. A. J. Salinas, A. I. Martín and M. Vallet-Regí, *J. Biomed. Mater. Res.*, 2002, **61**, 524.

203. M. Vallet-Regí, A. J. Salinas, J. Ramírez-Castellanos and J. M. González-Calbet, *Chem. Mater.*, 2005, **17**, 1874.

204. C. T. Kresge, M. E. Loenowicz, W. J. Roth, J. C. Vartuli and J. S. Beck, *Nature*, 1992, **359**, 710.

205. J. S. Beck, J. C. Vartuli, W. J. Roth, M. E. Loenowicz, C. T. Kresge, K. D. Schmitt, C. T. W. Chu, D. H. Olson, E. W. Sheppard, S. B. McCullen, J. B. Higgins and J. L. Schlenker, *J. Am. Chem. Soc.*, 1992, **114**, 10834.

206. F. Balas, M. Manzano, P. Horcajada and M. Vallet-Regí, *J. Am. Chem. Soc.*, 2006, **128**, 8116.

207. M. Vallet-Regí, F. Balas and D. Arcos, *Angew. Chem. Int. Ed.*, 2007, **46**, 7548.

208. P. Horcajada, A. Rámila, K. Boulahya, J. González-Calbet and M. Vallet-Regí, *Solid State Sci.*, 2004, **6**, 1295.

209. M. Vallet-Regí, I. Izquierdo-Barba, A. Rámila, J. Pérez-Pariente, F. Babonneau and J. M. González-Calbet, *Solid State. Sci.*, 2005, **7**, 233.

210. M. Vallet-Regí, L. Ruiz-González, I. Izquierdo-Barba and J. M. González-Calbet, *J. Mater. Chem.*, 2006, **16**, 23.

211. I. Izquierdo-Barba, L. Ruiz-González, J. C. Doadrio, J. M. González-Calbet and M. Vallet-Regí, *Solid State Sci.*, 2005, **7**, 983.

212. J. M. Gomez-Vega, M. Iyoshi, K. M. Kim, A. Hozumi, H. Sugimura and O. Takai, *Thin Solids Films*, 2001, **398–399**, 615.

213. X. Yan, C. Z. Yu, X. F. Zhou, J. W. Tang and D. Y. Zhao, *Angew. Chem. Int. Ed.*, 2004, **43**, 5980.

214. C. J. Brinker, Y. F. Lu, A. Sellinger and H. Y. Fan, *Adv. Mater.*, 1999, **11**, 579.

215. X. X. Yan, H. X. Den, X. H. Huang, G. Q. Lu, S. Z. Qiao, D. Y. Zhao and C. Z. Yu, *J. Non-Cryst. Solids*, 2005, **351**, 3209.

216. A. López-Noriega, D. Arcos, I. Izquierdo-Barba, Y. Sakamoto, O. Terasaki and M. Vallet-Regí, *Chem. Mater.*, 2006, **18**, 3137.

217. I. Izquierdo-Barba, D. Arcos, Y. Sakamoto, O. Terasaki, A, López-Noriega, and M. Vallet-Regí, *Chem. Mater.* 2008 (in press).

218. R. Z. LeGeros, *Prog. Cryst. Growth Charact.*, 1981, **4**, 1.

219. R. Z. LeGeros, J. P. LeGeros, O. R. Trantz and E. Klein, *Dev. Appl. Spectrosc.*, 1970, **7B**, 13.

220. H. Schmidt, *J. Non-Cryst. Solids*, 1985, **73**, 681.

221. H. H. Huang, B. Orler and G. L. Wilkes, *Polym. Bull.*, 1985, **14**, 557.

222. J. D. Mackenzie, Y. J. Chung and Y. Hu, *J Non-Cryst. Solids*, 1992, **147/148**, 271.

223. J. D. Mackenzie, *J. Sol-Gel Sci. Tech.*, 1994, **2**, 81.

224. S. Motakef, T. Suratwala, R. L. Poncone, J. M. Boulton, G. Teowee and D. R. Uhlmann, *J. Non-Cryst. Solids*, 1994, **178**, 37.

225. K. Tsuru, C. Ohtsuki, A. Osaka, T. Iwamoto and J. D. Mackenzie, *J. Mater. Sci.: Mater. Med.*, 1997, **8**, 157.

226. Q. Chen, F. Miyaji, T. Kokubo and T. Nakamura, *Biomaterials*, 1999, **20**, 1127.
227. C. Sanchez, B. Lebeau, F. Chaput and J. P. Boilot, *Adv. Mater.*, 2003, **15**, 1969.
228. C. Sanchez, B. Julián, P. Belleville and M. Popall, *J. Mater. Chem.*, 2005, **15**, 3559.
229. H. Schmidt, A. Kaiser, H. Patzelt and H. Sholze, *J. Phys.*, 1982, **12**, 275.
230. J. Livage, M. Henry and C. Sanchez, *Prog. Solid State Chem.*, 1988, **18**, 259.
231. C. Sanchez and F. Ribot, *New J. Chem.*, 1994, **18**, 1007.
232. A. I. Martín, A. J. Salinas and M. Vallet-Regí, *J. Eur. Ceram. Soc.*, 2005, **25**, 3533.
233. A. J. Salinas, J. M. Merino, N. Hijón, A. I. Martín and M. Vallet-Regí, *Key Eng. Mater.*, 2004, **254–256**, 481.
234. H. Schmidt and B. Seiferling, *Mater. Res. Soc. Symp. Proc.*, 1986, **73**, 739.
235. Y. Wei, D. Jin, "A new class of organic-inorganic hybrid dental materials" in Abstracts of papers of the American Chemical Society 214: 145-POLY Part 2, Sep. 7, 1997.
236. J. M. Yang, C. S. Lu, Y. G. Hsu and C. H. Shih, *J. Biomed. Mater. Res: Appl Biomater.*, 1997, **38**, 143.
237. S. Rhee and J. Choi, *J. Am. Ceram. Soc.*, 2002, **85**, 1318.
238. S. Yamamoto, T. Miyamoto, T. Kokubo and T. Nakamura, *Polym. Bull.*, 1998, **40**, 243.
239. Y. Hu and J. D. Mackenzie, *J. Mater. Sci.*, 1992, **27**, 4415.
240. Y. J. Chung, S. Ting and J. D. Mackenzie, in *Better Ceramics Through Chemistry IV*, vol. 180, B. J. J. Zelinski, C. J. Brinker, D. E. Clark and D. R. Ulrich eds., Materials Research Society, Pittsburg, PA, 1990, p. 981.
241. Q. Chen, F. Miyaji, T. Kokubo and T. Nakamura, *Biomaterials*, 1999, **20**, 1127.
242. N. Miyata, K. Fuke, Q. Chen, M. Kawashita, T. Kokubo and T. Nakamura, *J. Ceram. Soc. Jpn.*, 2003, **111**, 555.
243. Q. Chen, N. Miyata, T. Kokubo and T. Nakamura, *J. Mater. Sci.: Mater. Med.*, 2001, **12**, 515.
244. M. Kamitakahara, M. Kawashita, N. Miyata, T. Kokubo and T. Nakamura, *J. Mater. Sci.: Mater. Med.*, 2002, **13**, 1015.
245. M. J. Michalczyk and K. G. Sharp, US Patent 5,378,790, 1995.
246. K. G. Sharp and M. J. Michalczyk, *J. Sol-Gel Sci. Technol.*, 1997, **8**, 541.
247. K. G. Sharp, *Adv. Mater.*, 1998, **10**, 1243.
248. K. G. Sharp, *J. Mater. Chem.*, 2005, **15**, 3812.
249. M. Manzano, D. Arcos, M. Rodríguez Delgado, E. Ruiz, F. J. Gil and M. Vallet-Regí, *Chem. Mater.*, 2006, **18**, 5696.

CHAPTER 4

Clinical Applications of Apatite-Derived Nanoceramics

4.1 Introduction

In recent years, the development of nanotechnologies has acquired great scientific interest as they are a bridge between bulk materials and atomic or molecular structures. The properties of materials change as their size approaches the nanoscale and as the percentage of atoms at the surface of a material becomes significant. For bulk materials larger than one micrometre the percentage of atoms at the surface is minuscule relative to the total number of atoms of the material. The interesting and sometimes unexpected properties of nanoparticles are partly due to the aspects of the surface of the material dominating the properties in lieu of the bulk properties. The inherent nanoceramic properties allow tackling traditional problems of the bioceramics field, such as mechanical performance, bone-regeneration kinetics, biocompatibility, *etc.* as well as new challenges such as the optimisation of scaffolds for bone-tissue engineering and the design of nanodrug-delivery systems aimed to work within the bone tissue.

The nanoceramics contribution to biomaterials field is also mainly justified by their surface features.[1] It must be highlighted that the final aim is the optimum tissue–implant interaction, which is a surface event. The large surface area of nanoceramics supply new reactivity features. This fact involves new expectations for events such as bioactivity, bioresorption, foreign-body responses, *etc.* Secondly, nanoceramics give the chance to tailor at the nanometric scale the interactions between the material and the osteoblast adhesion proteins, with the purpose of optimising scaffolds for bone-tissue engineering. The nanoceramics surfaces are suitable to be easily functionalised and can incorporate biologically active molecules.[2] Since nanomaterials exhibit a maximum surface/volume ratio they are excellent candidates as vehicles for *drug-delivery* applications.

RSC Nanoscience & Nanotechnology
Biomimetic Nanoceramics in Clinical Use: From Materials to Applications
By María Vallet-Regí and Daniel Arcos

One of the classical drawbacks of bioceramics for orthopaedic applications is the mechanical behaviour when implanted in high-load body locations. The brittleness of these compounds has limited the clinical applications to the filling of small bone defects within load-bearing sites. Among the bioceramics with biomimetic properties, only some glass-ceramics, for instance A-W glass-ceramic[3] have evidenced a good performance in spine and hip surgery of patients with extensive lesions or bone defects due to its excellent mechanical strengths and capacity of binding to living bone. Glass-ceramics are defined as polycrystalline solids prepared by the controlled crystallisation of glasses. For instance, in the case of A-W glass-ceramic, apatite (38%) and wollastonite (34%) are homogenously dispersed in a glassy matrix (MgO 16.6, CaO_2 4.2, and SiO_2 59.2 wt%), taking the shape of a rice grain of 50 to 100 nm in size. Probably, this fact has inspired the use of nanoparticles to obtain highly resistant *ceramic/polymer nanocomposites*, which is currently one of the main topics in the nanostructured biomaterials field.

In addition to powder and blocks (both dense and porous) HA (hydroxyapatite) has been prepared for a long time at the micrometre scale as *coatings*. During recent years, significant research efforts have been devoted to nanostructure processing of HA coatings in order to obtain high surface area and ultrafine structure, which are properties essential for cell–substrate interaction upon implantation. The potential of nanoapatites as implant coatings has generated a considerable interest due to their superior biocompatibility, osteoconductivity, bioactivity, and noninflammatory nature.[4] One of the significant properties attributed to nanomaterials, namely, high surface area reactivity can be exploited to improve the interfaces between cells and implants. In addition, nanocrystallised characteristics have proven to be of superior biological efficiency. For example, compared to conventionally crystallised HA, nanocrystallised HA promotes osteoblast adhesion, differentiation and proliferation, osteointegration, and deposition of calcium-containing mineral on its surface, thereby enhancing the formation of new bone within a short period.[5]

Finally, the synthesis of hydroxyapatite nanopowders is also considered to improve the sintering processes of conventional ceramic bone implants. Sintering and densification of any ceramic depends on the powder properties such as particle size, distribution and morphology and it is believed that nanostructured calcium phosphate ceramics can improve the sintering kinetics due to a higher surface area and subsequently improve mechanical properties significantly.[6] Figure 4.1 shows some of the clinical and potential applications of nanoceramics in the field of bone grafting.

4.2 Nanoceramics for Bone-Tissue Regeneration

One of the most important aims of the nanoceramics with biomimetic properties is to provide new and effective therapies for those pathologies requiring bone regeneration. Among these diseases, osteoporosis must be highlighted due to the present and future high incidence within the world population.

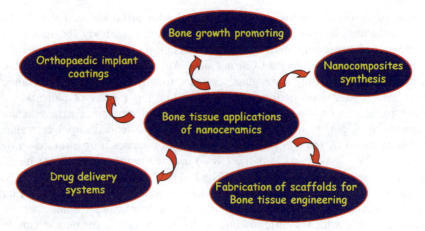

Figure 4.1   Current clinical applications of nanoceramics for bone-tissue repairing.

Osteoporosis affects 75 million people in Europe, USA and Japan[7] and 30–50% of women and 15–30% of men will suffer a fracture related to osteoporosis in their lifetime.[8] Pharmacological prevention and treatment of osteoporosis is the best strategy to date, although these therapies show some drawbacks related with bone formation in areas different from the osteoporosis sites. This fact is assumed as a consequence of the drug intake through systemic administration (oral and parentheral). In the case of fractures that cannot self-heal orthopaedic devices such as fixation devices or total hip prostheses are required. These devices have a limited average life time of around 15 years and it is speculated that this situation is due to lack of biomimetism that the implant surface exhibits at the nanometric scale.

Nanometrical calcium-phosphate-based biomaterials are very promising materials for both delivering drugs (as will be seen in Section 4.5) and for increasing bone mass. Through the implantation of calcium phosphates with properties mimicking the natural bone mineral, it is expected that properties such as bioactivity, dissolution range, resorption, *etc.* will be close to those of natural bones. Whereas previous emphasis was made to control the stoichiometry of these biomaterials, the aim of this last decade has been focused on controlling the size and morphology.[9] Actually, although bioceramics like HA and alumina with grain sizes greater than 100 nm are used for orthopaedic and dental implants because of its biocompatibility, these materials sometimes exhibit insufficient apposition of bone, leading to implant failure.[10–15]

Nowadays, there are experimental results that ceramics, metals, polymers, and composites with nanometre grain sizes stimulate osteoblast activity leading to more bone growth.[16] Long-term functions such as cell proliferation, synthesis of alkaline phosphatase and concentration of calcium in the extracellular matrix are enhanced when osteoblasts are seeded on nanoceramics.[17] Since there are chemical differences between osteoporotic and healthy bone,[18] calcium-phosphate-based nanoparticles can be formulated to selectively attach to areas of osteoporotic bone. The key point on the osteoblast selectivity is the protein adhesion at the first stages after implantation. Cells do not directly attach on materials surface, but on the proteins previously linked to the implant. Therefore, the first physical-chemical arguments to explain the better performance of nanoceramics, must be found in the events that occurred between the nanoceramics and the serum proteins.

4.2.1 Bone Cell Adhesion on Nanoceramics. The Role of the Proteins in the Specific Cell–Material Attachment

Proteins play a fundamental role in the bone cell adhesion. In fact, in the absence of serum proteins, the cell attachment to a substrate is dramatically decreased, whereas 10% of these proteins in a culture medium highly enhance the adhesion on ceramics with grain sizes below 100 nm.[19,20]

It is difficult to draw conclusions regarding optimal osteoblast adhesion as a function of the type of ceramic, since different grain sizes of several bioceramics such as alumina, titania, and HA have been tested, and enhanced osteoblast adhesion is observed on the three of them when exhibiting grain sizes below 100 nm.[21] Therefore, cellular responses to nanophase ceramics are independent of surface chemistry, at least among the biocompatible ceramics mentioned above.

Nanosized grains provide higher surface roughness in the range of tens of nanometres, which appears to be a critical characteristic that determines the nanoceramic biocompatibility. Moreover, the nanostructure provides a higher number of grain boundaries as well as an increased surface wettability, which is also associated with enhanced protein adsorption and cell adhesion. However, the enhanced biocompatibility of nanoceramics exhibits a much more interesting selective mechanism. When considering several protein anchorage-depending cells, for instance: osteoblast, fibroblast and endothelial cells, it is possible to correlate the adsorbed protein type and concentration with the observed cell adhesion on the materials tested. Figure 4.2 summarises this mechanism, where *vitronectine* is mainly adsorbed on nanoceramics, whereas *laminin* is preferentially adsorbed on conventional ceramics. Although the mechanism is not well established, fibroblast and endothelial cells attach preferentially on conventional ceramics, whereas osteoblasts are mainly adhered on nanoceramics.

The fact that nanophase ceramics adsorb greater concentrations of vitronectin while conventional ceramics adsorb greater concentrations of laminin explains the subsequent enhanced osteoblasts and endothelial cells adhesion on nanophase ceramics and conventional ceramics, respectively. Adhesion to substrate surfaces is imperative for subsequent functions of anchorage-dependent cells. Now the question is why vitronectine is mainly immobilised on nanoceramics, whereas laminin is more likely to attach to conventional ceramics. Webster *et al.*[21] explain this in terms of the inherent defects sizes of each kind of bioceramic. In addition to enhanced surface wettability, the roughness dictated by grain and pore size of nanophase ceramics influences interactions (such as adsorption and/or configuration/bioactivity) of determined serum

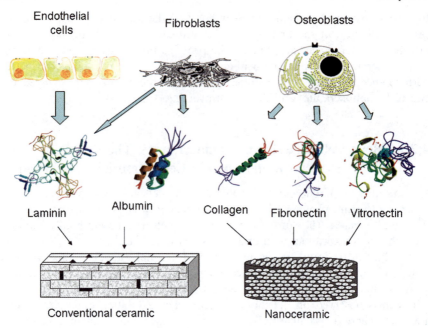

Figure 4.2 Schematic mechanism that explains the ceramic–protein–cell attachment specificity.

proteins and thus affect subsequent cell adhesion. In this sense, vitronectin, which is a linear protein 15 nm in length[22] may preferentially have adsorbed to the small pores present in nanophase ceramics, while laminin (cruciform configuration, 70 nm both in length and width) would be preferentially adsorbed into the large pores present in conventional ceramics.

The specificity of nanoceramics with respect to the type of cell has been also observed with bone marrow mesenchymal stem cells and osteosarcoma cells.[23] When both cultures are exposed to hydroxyapatite nanoparticles between 20 to 80 nm in diameter, greater cell viability and proliferation of mesenchymal stem cells were observed on the nano-HA, especially in the case of the smallest nanoparticles. On the contrary, the growth of osteosarcoma cells was inhibited by the nano-HA and the smallest particles exhibit the higher inhibitory effect.

Another example of the groundbreaking possibilities of nano-HA is the behaviour of the periodontal ligament cells in contact with this nanoceramic. In previous chapters, we have explained the role that HA plays in restoration of human hard tissue as well as in techniques that aim to regenerate periodontal tissues. Hydroxyapatite has osteoconductive effects but is nonbioresorbable and its use for periodontal tissue regeneration is not always effective. In fact, according to some studies, new periodontal regeneration is not always found if hydroxyapatite is used in the treatment of periodontal bone loss.[24] The key can be found in the poor response of the periodontal ligament cells to materials, which have only osteoconductive but no osteoinductive effect. Summarising, when hydroxyapatite is used in the periodontal osseous destruction, new periodontal regeneration is rarely found. This question has been tackled by implanting well-dispersed nano-HA powders. Appropriated particle dispersion can be achieved by using a sol-gel process in the presence of citric acid.[25] Citric acid acts as chelating reagent during the sol-gel process and prevents the agglomeration of hydroxyapatite. Nano-HA promotes the periodontal ligament cells (PDLC) proliferation as well as the alkaline phosphatase (ALP) activity. ALP plays a key role in the formation and calcification of hard tissues, and its expression and enzyme activity are frequently used as markers of osteoblastic cells. The high expression of ALP activity in the nanometre HA indicates that nanometre HA has the ability to induce osteogenic differentiation of PDLC. This fact points out that nano-HA may be a suitable grafting material for periodontal tissue regeneration.

4.2.2 Bioinspired Nanoapatites. Supramolecular Chemistry as a Tool for Better Bioceramics

Synthetic nano-HA with high levels of structural and chemical similarities with respect to those which occur in bone has been successfully synthesised during the last 20 years and this advance has been collected in several reviews.[26–28] However, the ability to prepare these compounds mimicking the morphological and organised complexity analogous to their biological counterparts has not yet been attained. We have seen that bone is a composite consisting of HA nanorods embedded in a collagen matrix. In this sense, it is thought that HA nanorods are desirable as building blocks for the long-range assembly of macroscopic biomaterials with hierarchical order, aimed to improve the implant's biocompatibility.[29] Organic matrix-mediated biomineralisation is a process that principally involves the use of organic macromolecular assemblies to control various key aspects of inorganic deposition from supersaturated biological solutions. In particular, the organic matrix plays an important role in delineating the structure and chemistry of the mineralisation environment, providing site-specific nucleation centres, regulating crystal growth and morphological expression, and facilitating the construction of higher-order assemblies.[30]

Significant progress has been made, for example, in crystal morphology using water soluble organic additives such as polyaspartic acid,[31,32] poly(acrylic acid),[33] and monosaccharides.[34] Similarly, ionic,[35–37] nonionic,[38] and block-copolymer[39] surfactants have been used to produce calcium phosphates with specific morphologies. In addition, self-assembled organic supramolecular structures have been employed as templates for the controlled deposition of calcium phosphate. This is the case of nano-HA synthesised within liposomes[40] or templated nano-HA synthesis within a collagen matrix leading to nano-HA/collagen composites.[41] All these possibilities are based on the fact that organised organic surfaces can control the nucleation of inorganic materials by geometric, electrostatic and stereochemical complementarities between the incipient nuclei and the functionalised substrates.[42–46]

By properly choosing organic additives that might have specific molecular complementarity with the inorganic component, the growth of inorganic nanocrystals can be rationally directed to yield products with desirable morphologies and/or hierarchical structures. Wang *et al.*[47] have prepared hydroxyapatite nanorods with tunable sizes, aspect ratios, and surface properties by properly tuning the interfaces between surfactants and the central atoms of HA based on the liquid–solid–solution strategy. This method is based on the phase transfer and separation process across the liquid, solid, and solution interfaces. By properly tuning the chemical reactions at the interfaces, an extensive group of nanocrystals with tunable sizes and hydrophobic surfaces has been prepared, demonstrating the effectiveness of controlling the chemical process occurring at the interfaces.

The preparation of HA nanorods in the presence of a cationic surfactant, cetyl trimethyl ammonium bromide (CTAB), has contributed to explain the formation mechanism of specific morphologies.[48] It is widely known that CTAB acts as a template,[49] with the template action resulting in the epitaxial growth of the product. Through the charge and stereochemistry features, molecule recognition occurs at the inorganic/organic interface.[34,50] The surfactant binds to certain faces of a crystal or to certain ions as well, so these ions are also incorporated to the existing nuclei at a steady rate and the final shape and size of HA particles can be well controlled.[51]

In the case of hydroxyapatite growth, the behaviour of CTAB is also considered to correlate with the charge and stereochemistry properties. In an aqueous system, CTAB ionises completely and results in a cation with tetrahedral structure, which can be well incorporated to the phosphate anion by the charge and structure complementarity. A probable mechanism for the templating process is that CTA^+-PO_4^{3-} mixtures form rod-like micelles, which contain many PO_4^{3-} groups on the surface, and when Ca^{2+} is added into the solution, $Ca_9(PO_4)_6$ clusters[52] are preferentially formed on the rod-shaped micellar surface due to conformation compatibility between identical hexagonal shape of the micelles and $Ca_9(PO_4)_6$ clusters.

The presence of a surfactant not only allows preparation of HA nanorods, but also can lead to the self-organisation to form ordered island-like bulk crystal complex structures through oriented attachment.[53] The conventional hydrothermal crystallisation process is a transformation process where amorphous fine nanoparticles act as the precursor. The formation of tiny crystalline nuclei in a supersaturated medium occurs at first and this is then followed by crystal growth. The large particles will grow at the expense of the small ones due to a higher solubility of the small particles than that of large particles. In the early stage, the examination of intermediate products shows the coexistence of the short rods, irregular nanoparticles and the longer nanorods.

However, in the presence of the self-assembled surfactant, the hydrothermal crystallisation process is limited in the controlled ordered space of water/surfactant interface. Initially, it is similar to the conventional hydrothermal crystallisation process. The formation of tiny crystalline nuclei in a supersaturated

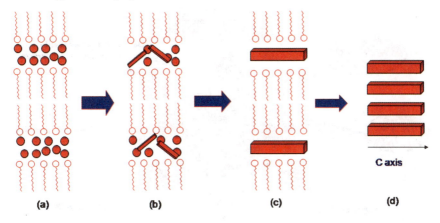

Figure 4.3 Formation process of oriented attachment of HA nanorods assisted by dodecylamine.

medium occurred at first and is then followed by crystal growth. But the difference is that intermediate products show the coexistence of the short rods and irregular nanoparticles formed only in the ordered limited space of water/surfactant interface. At the same time, while the small particles grow to long nanorods, the long nanorods are self-organised as building blocks through oriented attachment by sharing a common crystallographic orientation of HAP crystal and form island-like bulk crystals. This formation process of oriented attachment of nanorods can be schematically illustrated as in Figure 4.3.

4.3 Nanocomposites for Bone-Grafting Applications

In Chapter 2, we have described how HA is widely used for bone repair and tissue engineering due to its biocompatibility, osteoconductivity, and osteoinductivity. Through osteoconduction mechanisms, HA can form chemical bonds with living tissue. However, its poor biomechanical properties (brittle, low tensile strength, high elastic modulus, low fatigue strength, and low flexibility), when compared with natural hard tissues, limit its applications to components of small, unloaded, or low-loaded implants. One strategy to overcome this difficulty is to combine the bioactive ceramics with a ductile material, such as a polymer to produce composites. In recent years, the development of nanotechnology has shifted the *composite* synthesis towards the *nanocomposite* fabrication.

Nanocomposites are materials that are created by introducing nanoparticles (often referred to as *filler*) into a macroscopic sample material (often referred to as *matrix*). The main characteristic of nanocomposites is that the filler has at least one dimension in the range 1–100 nm. Currently, these materials constitute an important topic in the field of nanotechnology. Nanomaterial additives can provide very important advantages in comparison to both their conventional

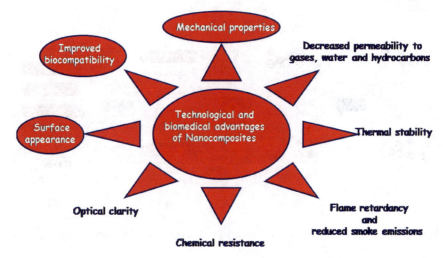

Figure 4.4   Advantages of nanocomposites respect to conventional composites. Those features related with biomaterials field are highlighted.

filler counterparts and base polymer. Figure 4.4 collects some of the most important advantages in materials science, highlighting the main properties with outstanding importance for bone-grafting applications: *mechanical properties* and *biocompatibility and surface features* improvement.

Several bioceramics have been used for the fabrication of nanocomposites. Among them we can highlight:

- Alumina (Al_2O_3)
- Zirconia (ZrO_2)
- Hydroxyapatite ($Ca_{10}(PO_4)_6(OH)_2$)

Alumina and zirconia belong to the first generation of bioceramics, characterised by an almost inert response after implantation and acceptable mechanical properties. Alumina has been used as a bearing couple in total hip replacement since the 1970s. As an artificial femoral head, alumina has demonstrated even better mechanical behaviour than metals, since its polished surface exhibit excellent wear resistance and produces less debris.[54] The osteoblast viability has been studied in the presence of nanosized alumina and titania particles, observing better cell proliferation independently of the chemical composition.[55] Several inorganic-inorganic nanocomposites such as alumina-zirconia and alumina-titania have been fabricated employing techniques like transformation-assisted consolidation and plasma spraying.[56] These combinations have resulted in nanocomposites with better fracture toughness and mechanical strength.

Zirconia exhibits chemical stability together with a good mechanical performance. For this reasons it has been used as a hard-tissue-repairing biomaterial. However, zirconia presents a similar drawback to alumina, *i.e.* ageing.

Degradation of zirconia is attributed to the transition of the tetragonal to monoclinic phase, followed by the occurrence of cracks from the surface to the inner bulk. Currently, yttria-stabilised zirconia (YSZ) is the preferred material for making ball heads.[57] Recently, hydroxyapatite/YSZ nanocomposites have been obtained with approximately 99% of the theoretical density.[58] These nanocomposites show improved mechanical properties (flexural strength and fracture toughness), which can be explained in terms of a uniform YSZ particle distribution in a nano-HA matrix that hinders the HA grain growth during the thermal treatment.

Although nanosized alumina, zirconia and titania can provide excellent mechanical properties as biomaterial components, none of them exhibit the biomimetic characteristic of nano-HA. Since this text is mainly devoted to nano-ceramics with biomimetic properties, special attention will be paid to composites formed by nano-HA as inorganic filler.

4.3.1 Nano-HA-Based Composites

Although nano-HA is an excellent artificial bone-graft substitute, its inherent low strength and fracture toughness have limited its use in certain orthopaedic applications. Fracture toughness of HA does not exceed $1.0\,\mathrm{MPa\,m^{1/2}}$ as compared to human bone ($2\text{--}12\,\mathrm{MPa\,m^{1/2}}$). Summarising, HA behaves as a typical brittle ceramic material[59] and HA-derived nanocomposites are an excellent alternative to overcome this problem. Compared with either pure polymers or conventional polymer composites, nanocomposites generally also exhibit an outstanding improvement in their mechanical properties.

From the point of view of the biological behaviour, nanocomposites promote an enhanced osteoblasts function as has been reported by Webster *et al.*[60] Besides conventional biopolymer composites studied[61–70], a number of investigations have recently been focused to determine the mineralisation, biocompatibility and mechanical properties of the nanocomposites based on various biopolymers. These groups of biocomposites mainly cover nano-HA/polylactide and its co-polymers,[71–74] nano-HA/chitosan,[75] nano-HA/collagen,[76–81] nano-HA/collagen/PLA,[82] nano-HA/gelatin[83–85] and the polycaprolactone semi-interpenetrating nanocomposites.[86]

In most of the cases, the improvement of the mechanical properties and the biological behaviour are the two main contributions provided by the apatite-derived nanocomposites.

4.3.2 Mechanical Properties of HA-Derived Nanocomposites

The incorporation of HA nanoparticles within a polymeric matrix leads to an increase of their mechanical parameters, mainly those related with the *dynamic mechanical properties* or *viscoelastic behaviour*. Dynamic mechanical analysis (DMA) is a technique used to study and characterise the viscoelastic nature of some materials, especially polymers. Two methods are currently used. One is the decay of *free oscillations* and the other is *forced oscillation*. Free-oscillation techniques involve applying a force to a sample and allowing it to oscillate after the force is removed. Forced oscillations involve the continued application of a force to the sample. An oscillating force is applied to a sample of material and the resulting displacement of the sample is measured. Since the sample deforms under the load, the stiffness of the sample can be determined, and the sample *storage* and *loss modulus* can be calculated. The storage and loss modulus in viscoelastic solids measure the stored energy (representing the elastic portion) and the energy dissipated as heat (representing the viscous portion), respectively. The tensile storage (E') and loss module (E'') are as follows:

$$E' = \frac{\sigma_0}{\varepsilon_0} \cos \delta \tag{4.1}$$

and

$$E'' = \frac{\sigma_0}{\varepsilon_0} \sin \delta \tag{4.2}$$

In these equations, σ (stress) and ε (strain) are defined as

$$\sigma = \sigma_0 \sin(t\omega + \delta) \tag{4.3}$$

and

$$\varepsilon = \varepsilon_0 \sin(t\omega) \tag{4.4}$$

where

ω is period of strain oscillation
t is time
δ is phase lag between stress and strain

From eqns (4.1) and (4.2), tan δ can be calculated, *i.e.* the ratio (E''/E'), which is useful for determining the occurrence of molecular mobility transition, such as the glass transition temperature.

The main dynamic mechanical effect of the nano-HA incorporation is the increase of the storage modulus with respect to the polymer and, of course, to the ceramic apatites. This means that nanocomposites exhibit higher elastic behaviour than their separated precursors. The storage modulus of nano-composites commonly increases with increased nano-HA content, indicating that hydroxyapatite has a strong reinforcing effect on the elastic properties of the polymer matrix. Since, the storage modulus reveals the capability of a material to store mechanical energy and resist deformation,[66] it can be stated that the higher the storage modulus, the more resistant the material is. The *loss modulus*, representing the ability to dissipate energy, also increases on raising the nano-HA content.

4.3.3 Nanoceramic Filler and Polymer Matrix Anchorage

A common problem with HA–polymer composites is the weak binding strength between the HA filler and the polymer matrix since they cannot form strong bonds during the mixing process. Often, the mechanical strength of the composite is compromised due to the phase separation of the HA filler from the polymer matrix. A clear example can be found in nano-HA/collagen nanocomposites. These compounds are one of the most studied systems because of their similarities with the natural bone. Actually, bone is an inorganic–organic composite material consisting mainly of collagen and HA, and its properties depend intimately on its nanoscale structures, which are dictated specifically by the collagen template.[87,88] Collagen is the major component of extracellular matrices, such as tendons, ligaments, skins and scar tissues in vertebrates.[89] However, the biocomposites of collagen and hydroxyapatite alone do not have adequate mechanical properties for various biomedical applications due to the weak filler–matrix interactions.

In order to improve the durability and mechanical properties of nano-HA/collagen composites, the use of polymeric binders has been proposed. Among these binders poly(vinyl alcohol), PVA, has shown a very good performance in terms of improving the filler/matrix binding[90] PVA hydrogels exhibit biocompatibility as well as a high elastic modulus even at relatively high water concentrations and have been employed in several biomedical applications including drug delivery, contact lenses, artificial organs, wound healing, cartilage, *etc.*[91,92] PVA has also been proposed as a promising biomaterial to replace diseased or damaged articular cartilage, but it has limited durability and does not adhere well to tissue.[93] However, the role that PVA can play as a binder between nano-HA and collagen fibres can be very interesting. The polar nature of PVA facilitates strong adhesion between the HAp and collagen. In this sense, nano-HA links to PVA through hydrogen bonding and by the formation of the $[OH]-Ca^{2+}-[OH]$ linkage, whereas the carbonyl groups of the collagen would be the active sites to bind to the nanocomposite components. The final result is an increase of the dynamic mechanical parameters, especially that related with the elastic properties (storage modulus) rather than the viscosity portion. Finally, the mechanical properties can be upgraded by cryogenic treatments, as has been already used in PVA hydrogel for heart-valve implant applications.[94,95] This effect is explained in terms of PVA crystallisation, which introduces strong interactions between different domains of the hydrogel.

Although the binder incorporation improves the mechanical performance of nanocomposites, the drawback of weak bonding of HA with polymers is still present, since they cannot form strong bonds during the mixing process. Another alternative to overcome this problem is coating the nano-HA with a polymer film. This coating must have functional groups able to form strong bonds with the polymer matrix. The polymer coating must be degradable so that the bioactivity of the nano-HA is not shielded.

For this purpose, Nichols *et al.*[96] proposed the radio-frequency plasma polymerisation technology to activate nano-HA powder surfaces by creating a degradable film with functional groups (*e.g.*, nano-HA-COOH) at nanoscale thicknesses. With this technique, the final biodegradability properties can be controlled through the experimental parameters such as RF power or gas pressure. For instance, under high-power and low-pressure conditions, the conversion of carboxyl groups into hydrocarbons, esters, or ketones/aldhydes is favourable, along with significant increases in crosslinking components, leading to nondegradable coatings. On the contrary, at low RF power, the crosslinking degree is minimised and the COOH retention on the coatings is high. Therefore, by using low plasma power in creating degradable coatings, fragmentation can be kept to a minimum and the functional groups can also be preserved from overpolymerisation.[97,98] The presence of these functional groups also provides active centres to improve the linkage between the nano-HA and the polymer matrix. As a consequence, the mechanical strength of the nano-HA–polymer scaffold is significantly improved with ultrathin degradable coatings when compared with uncoated control and nondegradable nanocoated groups.

As can be easily deduced, nanocomposite mechanical properties and bio-compatibility degree are also strongly dependent on the polymer used. The alkaline nature of HA often leads to a local pH increase of the environment. Moreover, the higher wettability and solubility of nanoparticles can result in higher pH increases that are harmful for the surrounding tissue. This problem can be partially overcome with the use of polymers with weak acid character. For instance, nanocomposites of nano-HA/polylactic acid and derived copolymers have provided very good results from the point of view of the biological response.

Not only the dynamic mechanical properties of biomaterials are enhanced by the nanoparticles incorporation. Parameters such as bending modulus have also been tested with different nanocomposites.[99] The bending module of nanocomposite samples of either poly(l-lactic acid) (PLA) or poly(methyl methacrylate) (PMMA) with 30, 40, and 50 wt% of nanophase (less than 100 nm) alumina, hydroxyapatite, or titania loadings were significantly greater than those of relevant composite formulations with conventional, coarser grained ceramics. The nanocomposite bending modules were 1–2 orders of magnitude larger than those of the homogeneous, respective polymer. Figure 4.5 clearly shows the mechanical improvement for three series of nanocomposites.

As can be seen in Figure 4.5 all of the nanoceramic/polymer composites exhibit increased bending moduli that are significantly greater than those of the corresponding conventional ceramic/polymer composites. It must be high-lighted that the bending moduli values for those nanocomposites with 40% by weight of nanoceramic content range between 1.0–3.5 GPa, *i.e.* in the range of 1–20 GPa exhibited by the human bone.[100] This increase in the strength of the nanophase ceramic/polymer composites, as compared to the conventional cer-amic/polymer composites, may be attributed to the fact that nanoparticles are better dispersed in the polymer matrix and the total interfacial area between the filler and matrix is higher for nanoscale fillers. Consequently, the nanoceramic

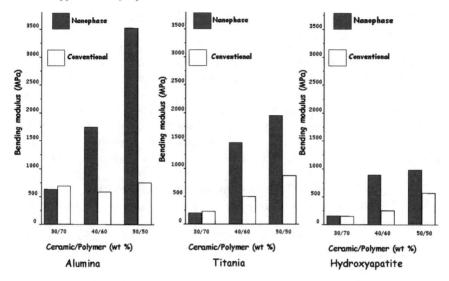

Figure 4.5 Bending moduli of ceramic/PLA substrates.

powders allow enhanced interactions between the filler and the matrix as compared to that of the conventional ceramic powders.

4.3.4 Significance of the Nanoparticle Dispersion Homogeneity

The importance of the nanocomposite synthesis strategy aiming to obtain a homogeneous nanoparticle dispersion has becomes a priority research line in this topic. Actually, most of the different procedures for nanocomposites fabrication are aimed to avoid this classical experimental problem, which is inherent to nanoparticles handling, *i.e.* the nanoparticles agglomeration. In order to overcome it, different strategies are available. For instance, ultrasonication stirring has been proven to be an effective strategy to avoid the agglomeration of particles in the polymer,[101,102] and good nanoparticle dispersion can be obtained when ultrasonication is combined with a solution casting method. After drying, nanocomposite films are obtained and subsequently shaped into the required geometry by hot pressing. In this way, Chen *et al.*[103] prepared nanocomposites based on bioresorbable polymer-poly(3-hydroxybutyrate-co-3-hydroxyvalerate) (PHBHV) by the incorporation of nano-HA using a solution casting method.

Kim[104] has proposed the use of an amphiphilic surfactant such as oleic acid to obtain nano-HA/poly(ε-caprolactone) (PCL) with the HA nanoparticles uniformly dispersed in the matrix. Oleic acid, which belongs to the fatty acid family and is generally noncytotoxic at low levels, mediates the interaction between the hydrophilic HA and hydrophobic PCL. With the mediation of oleic acid, the HA nanoparticles are distributed uniformly within the PCL matrix at the nanoscale.

4.3.5 Biocompatibility Behaviour of HA-Derived Nanocomposites

Nanoceramics in general and nano-HA in particular, are also incorporated to polymeric matrices to improve the cell–material interaction. Terms that explain how nano-HA improves the cell adhesion, proliferation and differentiation are analogous to those described in Section 3.1.3 for biomimetically grown nano-apatites. *In vivo* dissolution, adsorption of large amounts of serum proteins, increase of the surface roughness and ion dissolution signalling cells toward differentiation are upgraded features of nanocomposites when compared with conventional composites.

In 1998, Webster *et al.*[105] reported on the improved osteoblast adhesion on spherical nanosized alumina with grain size lower than 60 nm. In this work, a first precedent of the adhesion osteoblast selectivity was provided. Actually, whereas osteoblasts exhibit a better adhesion on nanosized ceramics, fibroblasts undergo an attachment decrease with respect to that observed for conventional ceramics. This fact has been subsequently corroborated on several nanoceramic/polymer nanocomposites[99] and is a very important advantage from the point of view of implant osteo-integration. Control of fibroblast function in apposition to orthopaedic/dental implants is desirable because fibroblasts have been implicated in the clinical failure of bone prostheses. Fibrous encapsulation and callus formation are the most frequently cited causes of incomplete osteointegration of orthopaedic and dental implants *in vivo*.[106–108] For these reasons, materials that have the desired cytocompatibility, *i.e.* that selectively enhance osteoblasts adhesion and subsequent functions of these cells, while at the same time minimising functions of competitive cells (such as fibroblasts), are very attractive.

When synthesising nanocomposites it must always be considered that cells do not directly attach to the material's surface, but to the adsorbed adhesion mediators proteins (fibronectin, vitronectin, laminin and collagen). Therefore, the immediate consequences of the new properties supplied by the nanosized fillers will be modifications on the protein adsorption. Together with higher surface area, nanoceramics introduce great changes in three aspects:

- *Surface defects and boundaries*, which lead to a reactivity increase in those sites where nanoparticles are placed.
- *Surface charge*. In this sense, HA doped with trivalent cations such as La^{3+}, Y^{3+}, In^{3+} or Bi^{3+} have shown better osteoblast adhesion, although these cationic substitutions involve a reduced grain size and it is difficult to differentiate between both concomitant effects.[109]
- *Surface morphology*. The dimensions of proteins that mediate cell adhesion and proliferation are at the nanometre level. Therefore, a surface with nanometre topography can increase the number of reaction sites compared to those materials with smooth surfaces.[21] The range of several tens of nanometres of surface roughness seems to be optimum for nanoceramic biocompatibility.

The enhancing of initial events during cell–biomaterial interactions (such as cell adhesion and concomitant morphology) clearly means an important advantage of nanoceramics incorporation into polymeric networks for dental and orthopaedic applications. However, evidence of long-term effects (cell proliferation, synthesis of alkaline phosphatase, and extracellular matrix mineralisation) is necessary before clinical use. These effects had been previously observed in ceramic surfaces modified with immobilised peptide sequences arginine-glycine-aspartic acid-serine (RGDS) and lysine-arginine-serine-arginine (KRSR)[110,111] contained in extracellular matrix proteins such as vitronectin and collagen. Nanoceramics are able to enhance these functions without peptide immobilisation,[17] exhibiting a much more efficient biological behaviour than conventional bioceramics.

4.3.6 Nanocomposite-Based Fibres

Nanocomposite fibrous structures are highly useful for the fabrication of porous biodegradable scaffolds. In this case, the homogeneity of nanoparticles dispersion becomes critical, and to develop the ceramic–polymer composite system as a micro-to-nanoscale structure, in the forms of fibres, tubes, wires, and spheres, the problem related to agglomeration and mixing needs to be primarily overcome.

Due to the high surface-area-to-volume ratio of the fibres and the high porosity on the submicrometre length scale of the obtained nonwoven mat, these materials have been proposed for biomedical applications,[112–115] including drug delivery, wound healing, and scaffolding for tissue engineering. The challenge in tissue engineering is the design of scaffolds that can mimic the structure and biological functions of the natural extracellular matrix (ECM).

Most of the work carried out to produce nanocomposite fibres has been through the electrostatic spinning (ES) technique. Electrostatic spinning or electrospinning is an interesting method for producing nonwoven fibres with diameters in the range of submicrometres down to nanometres. In this process, a continuous filament is drawn from a polymer solution or melt through a spinneret by high electrostatic forces and later deposited on a grounded conductive collector[116] as schemed in Figure 4.6. With this method electrospun fibres of nano-HA/polycaprolactone with different diameters have been obtained,[117] and used as scaffolding materials for the culture of preosteoblastic cells.[118]

Kim *et al.*[119] have also used this technique synthesising a nano-HA/PLA biocomposite system to produce fibrous structures. One of the main problems inherent to fibre fabrication is the difficulty in generating continuous and uniform fibres with the composite solution because of the innate problems of agglomeration. The incorporation of surfactants as a mediator between the hydrophilic HA and the hydrophobic PLA allows generating uniform fibres with diameters of 1–2 μm.

The electrostatic spinning technique has also been used to obtain nano-HA/PLA composites shaped as *membranes*, with application in bone-tissue

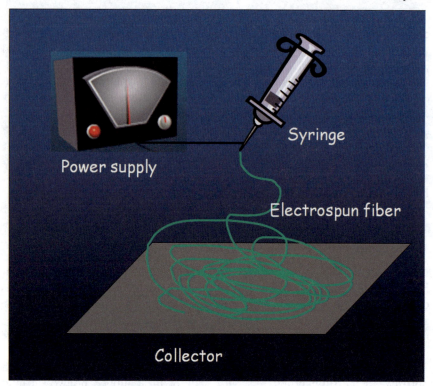

Figure 4.6   Scheme of an electrostatic spinning device.

regeneration.[120] The resulting membranes exhibit better cell adhesion and proliferation than the classical PLLA membranes, perhaps due to the constant pH of the environment when compared with PLLA degradation. As mentioned before, PLLA degradation results in a pH decrease that depending on the culture conditions can be harmful for osteoblasts. The presence of a soluble and slightly alkaline nano-HA buffers these changes and allows a better cell proliferation on the membranes. The mechanical properties such as the tensile strength, elastic modulus and strain to failure are highly improved, which can be explained in terms of a good HA dispersion that made the nanofibre matrix stiffer and less plastic in deformation, as could be expected from the incorporation of a hard inorganic phase.

4.3.7 Nanocomposite-Based Microspheres

Currently, much attention is focused on the fabrication of microspheres for biomedical application, with special significance as drug- and gene-delivery systems. Microspheres are widely accepted as delivery systems because they can be ingested or injected and present a homogeneous morphology.[121–124] Various approaches have been designed to prepare microspheres, depending on the chemical features of the final product. For instance, pyrolysis of an aerosol generated by ultrahigh-frequency spraying of the solution of precursors has been applied to prepare mesoporous silica microspheres encapsulating magnetic nanoparticles.[125] In the case of ceramic/polymer nanocomposites, for instance nano-HA/PLLA composites, microspheres are better prepared through methods involving oil-in-water emulsions.[72,126] The keystone of this strategy is the incorporation of a hydrophilic nanoceramic within the hydrophobic matrix. During the process, the inorganic particles tend to be located in water phase during the preparation process of the oil-in-water emulsion, thus a very small amount of inorganic particles was incorporated in the hydrophobic polymer microspheres. Qiu *et al.*[127] have proposed to functionalise the nano-HA surface with PLLA before carrying out the water-in-oil emulsion. With this strategy, these authors have prepared composite microspheres with uniform morphology and the encapsulated functionalised HA nanoparticles loading reached up to 40 wt% in the nano-HA/PLLA composite microspheres.

4.3.8 Nanocomposite Scaffolds for Bone-Tissue Engineering

Ceramic 3D porous scaffolds designed for bone-tissue engineering often show problems related with brittleness and difficulty of shaping. Ceramic/polymer composites can overcome these limitations, keeping the biocompatibility and bone-regenerative properties of some bioactive ceramics.[128–135] Among the main drawbacks of ceramic/polymer nanocomposites, we can highlight the organic solvents sometimes remaining in the composites and the coating of the ceramic by the polymer, which hinders its exposure to the scaffold surface. Figure 4.7 shows a clear example of this situation. Figure 4.7(a) shows the surface of a bioactive glass after being exposed to a biomimetic process in simulated body fluid. The surface appears fully covered by an apatite phase only one day after being treated with this fluid at 37 °C. On the contrary, Figure 4.7(b) shows the surface of the same bioglass as part of a composite with

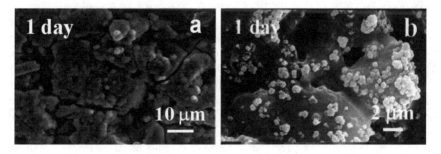

Figure 4.7   Surface of a bioglass (a) and a bioglass/PMMA composite (b) after one day in simulated body fluid.

poly(methyl methacrylate) (PMMA) after 1 day in SBF. The polymer skin that covers the ceramic is clearly seen and only separated apatite nuclei are observed, demonstrating an important delay of the biomimetic process.

In order to avoid these drawbacks poly(D,L-lactic-co-glycolic acid)/nano-hydroxyapatite (PLGA/HA) composite scaffolds have been fabricated by the gas-forming and particulate-leaching (GF/PL) method, without the use of organic solvents.[74] The GF/PL method exposed HA nanoparticles at the scaffold surface significantly more than the conventional solvent-casting and particulate-leaching (SC/PL) method does. The GF/PL scaffolds show interconnected porous structures without a skin layer and exhibit superior enhanced mechanical properties to those of scaffolds fabricated by the SC/PL method. The GF/PL method consists of shaping pieces of the corresponding polymer and ceramic together with NaCl. The conformed body is subsequently exposed to high-pressure CO_2 gas to saturate the polymer with the gas. Then, decreasing the gas pressure to ambient pressure creates a thermodynamic instability. This leads to the nucleation and growth of CO_2 pores within the polymer scaffolds. The NaCl particles are subsequently removed from the scaffolds by leaching the scaffolds in distilled water. With this strategy, highly porous PLGA/HA composite scaffolds can be fabricated exhibiting a higher exposure of HA at the scaffold surface and much better bone formation *in vitro* and *in vivo* than those fabricated by more conventional methods.

4.4   Nanostructured Biomimetic Coatings

The integration of any implant with bone tissue depends on the chemical and physical properties of the surface. In orthopaedic surgery, metals and their alloys are the most widely used implant materials due to their good mechanical properties, although in contact with body fluids or tissues they corrode.[136] An interesting alternative for protection of metal surfaces against corrosion is to coat the metal surface with a ceramic, which can act as an interface between the substrate and the bone, favouring the bone bonding. In this sense the calcium phosphates, such as HA and β-tricalcium phosphate (β-TCP), are common examples of such coatings.[137–139]

Nowadays, the most frequently employed technique to prepare commercial covered implants is plasma spraying.[140] However, this technique exhibits some disadvantages that cannot be easily avoided: unable to coat implants with complex shapes, differences in the chemical composition, delamination, *etc.* Other line-of-sight deposition methods such as sputtering[141] or laser ablation[142] do not solve, for instance, the problem of coating of porous substrates. Other physical methods include magnetron sputtering, ion-beam coating, anode oxidation and anodic spark deposition; extended information can be found in bibliography.[143–147] Chemical vapour deposition (CVD) techniques have been successfully used for the preparation of calcium phosphate coatings. By using this coating strategy, thin films of calcium phosphates can be obtained, where the microstructure, crystallinity and composition of the deposited films can be controlled by modifying the composition of the precursor solution, reactor atmosphere and substrate temperature.[148,149]

Solution-based methods are an emerging option for the preparation of these coatings due to several features: better control of coating morphology, chemistry and structure, covering of intricate pieces, simplicity of technology, *etc.* In this section we will mainly deal with *sol-gel* and *biomimetic deposition coating* procedures, two strategies that lead to highly biocompatible nanostructured coatings with a wide range of possibilities to incorporate therapeutic agents, growth factors, adhesion proteins, peptides sequences, *etc.* due to the low-temperature processes involved in these methods.

4.4.1 Sol-Gel-Based Nano-HA Coatings

The *sol-gel* process[150,151] is a wet-chemical technique for the fabrication of materials (typically a metal oxide) starting from a chemical solution that reacts to produce colloidal particles (*sol*). Typical precursors are metal alkoxides and metal chlorides, which undergo hydrolysis and polycondensation reactions to form a colloid, a system composed of solid particles (size ranging from 1 nm to 1 μm) dispersed in a solvent. The sol then evolves towards the formation of an inorganic network containing a liquid phase (*gel*). Formation of a metal oxide involves connecting the metal centres with oxo (M–O–M) or hydroxo (M–OH–M) bridges, therefore generating metal-oxo or metal-hydroxo polymers in solution. The *drying* process serves to remove the liquid phase from the gel, thus forming a porous material, and then a thermal treatment (*firing*) may be performed in order to favour further polycondensation and enhance mechanical properties.

The precursor sol can be either deposited on a substrate to form a film (*e.g.* by dip-coating or spin-coating), cast into a suitable container with the desired shape (*e.g.* to obtain monolithic ceramics, glasses, fibres, membranes, aerogels), or used to synthesise powders (*e.g.* microspheres, nanospheres). The sol-gel approach is interesting in that it is a cheap and low-temperature technique that allows the incorporation of drugs and osteogenic agents within the coatings. Hijón *et al.*[152] prepared bioactive nano-HA coatings deposited on Ti6Al4V by the sol-gel dipping technique (see Figure 4.8), from aqueous solutions containing triethyl phosphite and calcium nitrate, although other precursors can be also used to prepare HA coatings by the sol-gel techniques, as shown in Table 4.1.

Nano-HA coatings with particle sizes of around 75 nm and controlled roughness can be prepared by modifying the drying temperature in the range of 30–60 °C. A decrease in the R value is observed as the ageing temperature increases as can be seen from the roughness profiles obtained by scanning force microscopy shown in Figure 4.9. The R values obtained were 11, 8 and 5 nm for layers dried at 30, 40 and 60 °C, respectively. The coating thickness is around 0.2 μm per dipping cycle. To obtain HA coatings with higher thickness, the dip-coating method is repeated several times (up to 10 times). In coatings of six or more layers, the formation of cracks on the coating is likely to occur, and a

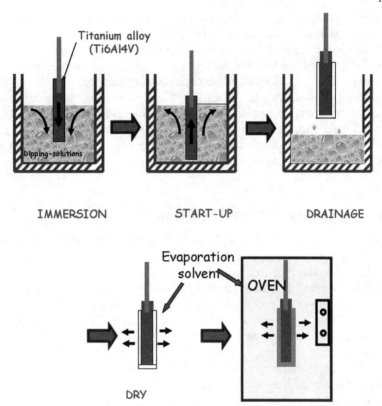

Figure 4.8 Scheme followed for the preparation of coatings by the dip-coating process.

valid compromise between the coating thickness and integrity can be reached for six layers, *i.e.* 5 or 6 dipping cycles. In general, the coating roughness decreases as more layers are incorporated to the coating.

Another factor influencing the final coating composition, textural properties and homogeneity is the water presence in the sol. Actually, the precursors:H_2O ratio determines the hydrolysis/polycondensation kinetic and the final coating characteristics. In general, sols containing higher amounts of ethanol require longer ageing times and lead to purer HA as well as to more homogeneous coatings.[170] These coatings exhibit tensile strength adhesion values of around 20 MPa[171] comparable to those obtained for HA coatings prepared by other coating strategies, such as electrodeposition, plasma spray or pulsed laser deposition.[142,172,173]

Nano-HA coatings prepared by sol-gel dipping exhibit biomimetic behaviour when they are exposed to simulated body fluid at 37 °C. Contrary to other CaP-based bulk materials, the development of a new carbonated calcium-deficient apatite phase can be observed by the nucleation and growth of biomimetic crystals that are observable by SEM. Figure 4.10 shows this new apatite-like

Table 4.1 Some of calcium and phosphorous precursors used in the synthesis of HA coatings deposited by sol-gel technique.

Reference	Calcium precursor/ solvent	Phosphorous precursor/solvent
Brendel et al.[153]	Calcium nitrate/acetone	Phenyldichlorophosphine/acetone/ water
Russell et al.[154]	Calcium nitrate/ 2 methoxyethanol	N-Butyl acid phosphate/ 2 methoxyethanol
Hsieh et al.[155]	Calcium nitrate/ 2 methoxyethanol	Triethyl phosphate/2 methoxyethanol
Goins et al.[156]	Calcium nitrate/ 2 methoxyethanol	Diethyl phosphite/2 methoxyethanol
You and Kim[157]	Calcium nitrate/ methanol	Triethyl phosphite/methanol
Hwang and Lim[158]	Calcium nitrate/ methanol	Phosphoric acid/methanol
Kojima et al.[159]	Calcium nitrate/ethanol	Triethyl phosphate/ethanol
Liu et al.[160]	Calcium nitrate/ethanol	Triethyl phosphite/ethanol/water
Gan and Pilliar[161]	Calcium nitrate/ethanol	Triethyl phosphite/water Ammonium dihydrogen phosphate/ water
Piveteau et al.[162]	Calcium nitrate/ethanol	Phosphoric pentoxide/ethanol
Cavalli et al.[163]	Calcium nitrate/water	Diammonium hydrogen phosphate/ water
Weng and Baptista[164]	Calcium glycoxide/ ethyleneglycol	Phosphoric pentoxide/ethanol or butanol
Chai et al.[165]	Calcium diethoxide/ ethanol/ethanediol	Triethyl phosphite/ethanol
Gross et al.[166]	Calcium diethoxide/ ethanol/ethanediol	Triethyl phosphite/ethanol/ ethanediol
Haddow et al.[167]	Calcium diethoxide/ ethanediol	Triethyl phosphite/ethanediol
Ben-Nissan et al.[168]	Calcium diethoxide or calcium acetate/ethyle-neglycol/acetic acid	Diethylhydrogenphosphonate
Tkalcec et al.[169]	Calcium 2-ethylhex-anoate/ethylhexanoic acid	2-ethylhexylphosphate/ ethylhexanoic acid

layer on a HA coating, which exhibits similar morphology to those biomimetic layers appearing on the surface of other highly bioactive bioceramics. TEM observation (Figures 4.10(c) and (d)) demonstrates the different morphology of the nano-HA particles that constitute the coating, and the nano-HA forming the new biomimetic layer.

The crystals corresponding to the sol-gel coating (Figure 4.10(c)) are nano-sized (<25 nm), round-shaped and with a Ca/P molar ratio of 1.7 ± 0.1, according to the EDS spectra. The ones formed into the SBF solution (Figure 4.10(d)) are larger and show a needle-like shape; the Ca/P molar ratio of this kind of crystals was found to be 1.4 ± 0.1, similar to the ratio observed in biological apatites[174] and other apatites formed in SBF.[175] In the same way, both ED

Figure 4.9   SFM 3D image corresponding to the nano-HA coating deposited from sols aged at different temperatures. (a) One layer dried at 30 °C ($R = 11$ nm). (b) One layer dried at 40 °C ($R = 8$ nm), (c) one layer dried at 60 °C ($R = 5$ nm) and (d) six layers dried at 60 °C ($R = 4$ nm).

patterns show diffraction rings that can be indexed to the interplanar spacings of an apatite phase. In addition, the ED diagram in Figure 4.10(c) also shows diffraction maxima that are indicative of the higher crystallinity of sol-gel-derived HA nanocrystals when compared to those obtained in SBF.

The bioactive behaviour of the nano-HA coatings can be improved by silicon incorporation into the apatite structure.[176] Through the incorporation of stoichiometric amounts of tetraethyl orthosilane, $Si(OCH_2CH_3)$ within the sol, silicon-substituted hydroxyapatite coatings, according to the formula: $Ca_{10}(PO_4)_{6-x}(SiO_4)_x(OH)_{2-x}\mu_x$ where x varies from 0.25 to 1 and μ expresses the anionic vacancies generated. The presence of carbonated species in the sol led to a final coating composition with general formula $Ca_{10}(PO_4)_{6-x-y}(SiO_4)_x(CO_3)_y(OH)_{2-x+y}$ where carbonates are included in the phosphate sites competing with the introduced silicates.

Surface sol-gel processing, a variant of the bulk sol-gel dip-coating method, can be used to fabricate ultrathin metallic oxides with nanometre precise control.[177] The layer-by-layer process begins with the chemisorption of a hydroxyl-functionalised surface in a metal alkoxide solution followed by rinsing, hydrolysis, and drying of the film. This sol-gel reaction occurs on the surface of the substrate each time the hydroxyl groups, TiOOH, are regenerated to form a monolayer of TiO_2 and repetition of the entire process results in

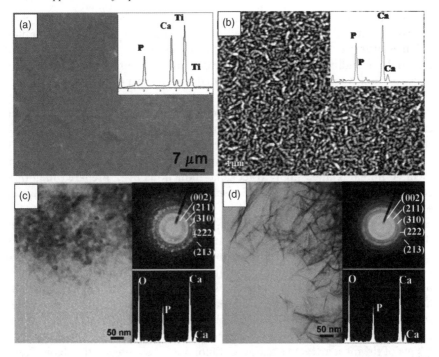

Figure 4.10 Biomimetic behaviour of apatite nanocoatings. (a) Scanning electron micrograph and EDX spectrum of a Ti4AlV substrate coated by a nano-HA phase. (b) Scanning electron micrograph and EDX spectrum of a Ti4AlV substrate coated by a nano-HA phase after 7 days in SBF. (c) Transmission electron microscopy image, electron diffraction pattern and EDX spectrum of nano-HA sol-gel coating. (d) Transmission electron microscopy image, electron diffraction pattern and EDX spectrum of biomimetic apatite growth after 7 days in SBF.

multilayers of the thin oxide film. A calcination or sintering process may be applied if a denser or more crystalline oxide is desired, but this is often unnecessary.[178] The process is readily applied to any hydroxylated surface, using a metal alkoxide reactive to OH groups, and the sol-gel procedure is independent of each cycle, which allows individual layers to be nanostructured.[179] With this strategy, coated Ti6Al4V substrates can be obtained with corrosion behaviour as good as TiO_2, but with increased bioactivity under biomimetic conditions.

4.4.2 Nano-HA Coatings Prepared by Biomimetic Deposition

In Chapter 3 it was seen how supersaturated solutions with ionic compositions similar to that of human plasma can be used with the aim of mimicking the mineralisation process. In that chapter, bioceramics that induce the growth of nanoapatites in contact with biomimetic solutions were considered as "nanoapatite producers" since their chemical composition and textural characteristics induced the nucleation and subsequent growth of bone-like apatites.

Supersaturated solutions can be also employed to coat complex-shaped materials, including metals. The most widely used biomimetic solution is SBF. This solution is just slightly supersaturated with respect to the precipitation of HA, and consequently the nucleation and precipitation processes on metal surfaces are quite slow. In order to accelerate the deposition, high ionic strength calcium phosphate solutions and pretreatment with highly supersaturated solutions, ($3\times$, $5\times$, or even $10 \times SBF$) can be used, as explained in Chapter 3.

Biomimetic nano-HA coatings greatly increase the osteoconductivity of metallic implants and supply osteogenic induction. Li[180] prepared nano-HA coatings on grit-blasted Ti6Al4V by soaking these specimens in a biomimetic solution highly concentrated in Ca^{2+} (6.0 mM) and HPO_4^{2-} (2.4 mM). The coating formation took place after 3 days at 45 °C and the differences between the coated and uncoated specimens were evident. When implanted in the distal femur of dogs, greater bone formation was generated in those surfaces lined with the apatite coating than those of the noncoated titanium surface. Human osteoblasts also exhibit clear differences when they are cultured in coated and noncoated Ti6Al4V substrates. The human osteoblasts cultured on coated substrates develop a more 3D morphology as well as a higher number of anchorage elements than those on noncoated surfaces.

One of the most attractive features of biomimetic coatings is that biologically active molecules, such as drugs, osteogenic agents, growth factors, *etc.* can be coprecipitated with the apatite crystals onto metal implants,[181,182] which can be subsequently released during the coating degradation acting as a drug-delivery system. The retardant effect of serum albumin (SA) on the biomimetic nano-HA formation is well known and was explained in Chapter 3. However, under high Ca^{2+}, PO_4^{3-} or HPO_4^{2-} concentration conditions the effect of (SA) is reflected as changes in the crystallite morphology, but not as an inhibitory effect. The nano-HA crystallites decrease in size, assume a marked curvature and become more densely packed as a function of SA concentration in the solution.[183]

Coprecipitation of active agents with biomimetic nanocoatings also provides a very important advantage with regard to the kinetic release. Although the use of apatite nanoparticles as drug-delivery systems will be considered in the next section, it is important to highlight the effect of the coprecipitation methods compared with the adsorption onto preformed coatings. Most of the therapeutic agents adsorbed on preformed coating are released in a single fast burst effect. On the contrary, therapeutics incorporated by coprecipitation are gradually released over several days, enhancing their potential as controlled drug-delivery carriers.

Biomimetic coprecipitation methods allow the nucleation and growth of a variety of calcium phosphates. For instance, carbonate hydroxyapatites (CHA) or octacalcium phosphate (OCP) can be also precipitated as biomimetic coatings. CHA and OCP have different pH stability, solubility and show different resorption under the osteoclasts action.[184] Barrère *et al.*[185] have demonstrated that OCP biomimetic coatings exhibit higher osteogenic action when implanted in both intramuscular and bone locations. The presence of the Ca-P coating during an appropriated time period (*i.e.* coating stability) and the architecture of the implant seem to be very important conditions. Very fast coating dissolution and flat or dense surfaces do not induce an osteogenic response when implanted in intramuscular locations. Anyway, bone induced into muscular implantation is degraded with time. When Hedrocel™ cylinders (porous tantalum) are biomimetically coated with OCP and CHA and placed into bone tissue a direct bonding between the implant and the host bone occurs. However, only OCP-coated cylinders exhibit bone ingrowth in the center of the implant, although this new bone is not necessarily in contact with the host bone. This suggests that OCP coatings exhibit a higher osteogenic behaviour than CHA coatings. This difference in the osteogenic behaviour could be explained by the lower CHA coating in the bone environment. In this location, the resorption of the coating mainly depends on the osteoclastic activity. In this sense, osteoclastic activity is higher on biomimetic CHA, as had been previously demonstrated.[184] Moreover, the rougher surface provided by the larger and sharp vertical OCP crystals seems to provide a more appropriated microstructure to influence bone formation.

4.5 Nanoapatites for Diagnosis and Drug/Gene-Delivery Systems

4.5.1 Biomimetic Nanoapatites as Biological Probes

Biomedical probes possess interesting diagnosis properties as intracellular optical sensors.[186–188] Especially relevant are those luminescent probes that exhibit a fluorescent signal as a response, due to the high sensibility showed by fluorescence spectroscopy. A wide variety of fluorescent organic molecules are currently used as biological probes, which enable molecules in cells to be visualised by fluorescence. Although this method is sensitive, degradation of the organic molecule under irradiation, leads to a rapid fall in fluorescence intensity.

In this sense, the incorporation of quantum dots (QD)[189] to the field of diagnosis and therapy has meant a significant advance. QDs are inorganic fluorophores that exhibit size-tunable emission (*i.e.* there is a predictable relationship between the size of the QD and its emission wavelength), strong light absorbance, bright fluorescence, narrow symmetric emission bands, and high photostability QDs. The problem is that QD cores are usually composed of elements from groups II and VI, like CdSe, or groups III and V, *e.g.*, InP, while the shell is typically a high bandgap material such as ZnS.[190] Since cadmium and selenium can be highly toxic, the search for more compatible compounds, such as biomimetic apatites with luminescent properties, is a priority research line in the development of nanodiagnosis. In addition to the high biocompatibility, calcium phosphate nanoparticles might undergo long term dissolution inside the cells due to the lower Ca^{2+} concentration in the intra-cellular compartment.[191,192]

Doat *et al.*[193,194] have synthesised and studied biomimetic calcium-deficient apatite nanocrystallites doped with trivalent europium (Eu^{3+}). The composition and the crystallite sizes of such an apatite enable it to interact with living cells and therefore to be exploited as a biological probe. These apatites were synthesised at 37 °C by coprecipitating a mixture of Ca^{2+} and Eu^{3+} ions by phosphate ions in a water–ethanol medium. The nanoparticles are internalised by human epithelial cells and their luminescence stability allows them to be observed by confocal microscopy.

The required size-tunable properties of these probes dictate a size range of 2–6 nm, exhibiting dimensional similarity with biological macromolecules, *e.g.* nucleic acids and proteins. Most of the proposed synthetic nanoapatite routes commonly yield slightly aggregated bioapatite nanoparticles and individual-isation of the primary crystallites has to be achieved for a better spectral and spatial resolution in biological applications. In addition, in order to minimise the influence of the luminescent nanocrystals on the biological mechanisms, decreasing the size of the individualised nanoparticles in the range of small proteins or oligonucleotides is desirable. Lebugle *et al.*[195] have described the preparation of individualised monocrystalline colloidal apatitic calcium phosphate nanoparticles stabilised at neutral pH and using aminoethyl phosphate (NH_3^+–CH_2CH_2–PO_4H_2). This strategy has been successfully applied to the synthesis of various doped calcium phosphate nanoparticles. Doping with luminescent centres such as Eu^{3+}, Tb^{3+}, *etc.*, yields a range of calcium phosphate nanophosphors suitable for biological labelling. Finally, the colloidal stability in neutral pH must be achieved. For this aim, the use of functional amino surface groups, which offers the further possibility of bioconjugation, is a suitable strategy to achieve the appropriate stability.

4.5.2   Biomimetic Nanoapatites for Drug and Gene Delivery

Drug-delivery systems (DDSs) can be described as formulations that control the rate and period of drug delivery (*i.e.* time-release dosage) and target specific areas of the body. Currently, local drug delivery is an ever-evolving strategy that responds to the development of new active molecules and potential treatments, such as gene therapy. Actually, the research efforts in the pharmaceutical field are leading to the evolution of new therapeutic agents but also to the enhancement of the mechanisms to administrate them.[196–199]

The field of nanotechnology in recent years has motivated researchers to develop nanostructured materials for biomedical applications. In a similar way to silica-based mesoporous materials,[200–204] the biomimetic calcium phosphate nanoparticles for drug delivery have experienced outstanding advances in recent years. These nanosystems are especially promising for those pathological situations associated to bone surgery, such as bone-tumour extirpation, infection risk, acute inflammatory response, *etc.* Therefore, it can be stated that local drug release in bone tissue is one of the most promising therapies in orthopaedic surgery. Oral administration commonly requires very high and, sometimes, low effective dosages to reach high enough drug concentrations in the poorly irrigated bone tissue. Antibiotics, growth factors, chemotherapeutic agents, antistrogens and anti-inflammatory drugs are good candidates for the most common bone-related therapies.

4.5.2.1 Biomimetic Nanoapatites for Bone-Tumour Treatment

One the most promising therapeutic actions of drug-loaded biomimetic nanoapatites is the treatment of bone cancer. For instance, osteosarcomas and Ewing's sarcoma are malignant bone tumours commonly occurring in children's growing bones. No evolution of the survival rates has been recorded for two decades in response to current treatment, often associated with toxic and badly tolerated cures of chemotherapy with low therapeutic response.[205] Thus, treatment for these bone cancers commonly involves surgery, such as limb amputation, or limb-sparing surgery and consequently, the high loss of bone tissue is one of the main drawbacks after a bone-tumour extirpation. A second problem is that malignant cells can remain around the site, leading to tumour recurrence with fatal consequences. In this sense, the use of biomimetic calcium phosphate grafts combined with local specific cancer treatments is an excellent alternative to restore bone defects such as those that occur after tumour extirpation.

Cis-diaminedichloroplatinum (cisplatin) is one of the most active anticancer agents in the treatment of osteosarcoma, but must be used in limited short-term, high-dose treatments because of nephrotoxicity and ototoxicity. The minimisation of the systemic toxicity of chemotherapeutic drugs including cisplatin has been demonstrated after local intratumoral treatments with comparable antitumour efficacy to that of a systemic dose.[206–214] Cisplatin can be easily adsorbed onto hydroxyapatite nanoparticles.[215] The chemical and physical characteristics of the apatite crystals, including the chemical composition, structure, porosity, particle size and surface area, as well as the ionic composition of the equilibrating solution (pH, ionic strength, concentration of ion constituents), all play an important role in both the binding and release of the specific chemical components from calcium phosphates.[216–220]

Nanoapatites bind higher amounts of cisplatin than well-crystallised apatites. This linkage is achieved through an endothermic process, since the cisplatin immobilised onto the apatite surface is 3 times higher when the adsorption process is accomplished at 37 °C than when it is carried out at 24 °C.[215] Regarding the drug release, nanoapatites deliver cisplatin more slowly that the more crystalline apatites, although in both cases the presence of chloride ions in the surrounding fluid is needed for cisplatin release. The greater activity of the poorly crystalline apatites in the cisplatin adsorption process, compared to the well-crystallised hydroxyapatite, can be attributed to the presence of more surface defects (nonapatitic environments) that create active binding sites.

Moreover, the morphology and size of the crystals, other surface irregularities, and lower crystallinity also account for the higher reactivity of these materials. The cytotoxicity tests with these apatite/cisplatin systems demonstrate that these formulations exhibit cytotoxic effects on K8 cells with a dose-dependent decrease of the cell viability.

The adsorption and release of Pt-derived antitumoral drugs can be tuned by controlling the nanoparticle morphology of the biomimetic nano-HA. Actually, there exist several synthesis routes that allow tailoring the shape and surface composition of the nanoparticles. For instance, Palazzo *et al.*[221] have proposed the preparation of biomimetic nano-HA with both needle-shaped and plate-shaped morphologies and different physical-chemical properties. For instance, *needle-shaped* nanocrystals can be prepared from an aqueous suspension of $Ca(OH)_2$ by slow addition of H_3PO_4,[222,223] obtaining needle-shaped nanocrystals having a granular dimension around 100 ± 20 nm. Besides, *plate-shaped* nanocrystals can be precipitated from an aqueous solution of $(NH_4)_3PO_4$ by slow addition of an aqueous solution of $Ca(CH_3COO)_2$ keeping the pH at a constant value of 10 by addition of $(NH_4)OH$ solution. The reaction mixture is stirred at 37 °C for 72 h and then the deposited inorganic phase is isolated by filtration of the solution, repeatedly washed with water, and freeze-dried. In this way, *plate-shaped* nano-HA are obtained having granular dimensions of 25 ± 5 nm.[224] Although the bulk Ca/P ratios are similar in both kinds of nano-HA (between 1.65 and 1.62), the surface Ca/P ratio is lower for the *needle-shaped* nanocrystals (1.30) compared with the HAps particles (1.45), suggesting that the former is more surface deficient in calcium ions.

Taking into account the specific properties of each drug (negative, positive or neutral charge) and the nano-HA features specific nano-HA/drug conjugates can be tailored for specific clinical situations. The adsorption and desorption kinetics are dependent on the specific properties of the drugs and the morphology of the HA nanoparticles. In addition to cisplatin, di(ethylenedi-amineplatinum)medronate (DPM) has also been incorporated to biomimetic nano-HA (see Figure 4.11). DPM belong to the family of platinum(II) compounds with aminophosphonic acids and have also been proposed as a means for targeting cytotoxic moieties to the bone surface.[225,226] These compounds exhibit antimetastatic activity, reduce bone-tumour volume and are less nephrotoxic than cisplatin.[227–229]

Adsorption of the platinum complexes occurs with retention of the nitrogen ligands but the chloride ligands of cisplatin are displaced. Consequently, the positively charged aquated cisplatin is strongly adsorbed on the negatively charged nano-HA surface, while the neutral DPM complex shows lower affinity towards the negatively charged nanoceramic. Anyway, the adsorption of the two platinum complexes is driven by electrostatic attractions. Consequently, adsorption of positively charged hydrolysis species of cisplatin is more fa-voured on the phosphate-rich needle-shaped nano-HA surface, while the neu-tral DPM complex shows lower affinity for needle- or plate-shaped, negatively charged apatitic surfaces.

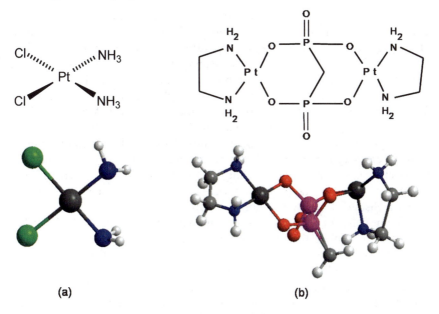

Figure 4.11 Molecular structure of (a) cis-diaminedichloridoplatinum (II) (cisplatin), (b) di(ethylenediamineplatinum)medronate (DPM).

This kind of short-range electrostatic interactions dominate the kinetic release and drug desorption is faster for neutral DPM than for charged aquated cisplatin. The release of DPM takes place through complete cleavage of the platinum–medronate bond and the release is greater when adsorbed to needle-shaped rather than plate-shaped nano-HA. As can be seen, these processes are modulated to some extent by the surface composition demonstrating that biomimetic nano-HA can be tailored for specific therapeutic applications.

4.5.2.2 Nanoapatites as Antibiotic-Delivery Systems

Nano-HA is used to improve the performance of polymeric antibiotic delivery systems through the formation of nanocomposites, which exhibit an enhanced drug-loading capacity as well as better release performance. For instance, nano-HA can be incorporated to polylactide-based systems[230] increasing the potential of PLA for biomedical applications in general and for drug delivery in particular. With this aim, several biodegradable polymers have been combined with nano-HA. For instance, Wang *et al.*[231] prepared polyhydroxybutyrate-co-hydroxyvalerate (PHBV)-nano-HA microparticles for gentamicin release. The idea is to fabricate a long-term drug-release system by preparing drug-loaded HA nanoparticles, followed by the encapsulation of nanoparticles with a bio-degradable polymer, such as PHBV. A classical problem of polymeric micro-spheres for drug delivery is that microspheres are generally prepared by double or single emulsion solvent evaporation methods. The outer phase, which usually was aqueous phase, induced hydrophilic drugs like gentamicin to move out of the polymer phase, resulting in comparatively lower encapsulation efficiency of gentamicin.[232,233] In the case of nanocomposites, a kind of hybrid structure consisting of HA nanoparticles and polymers can be fabricated to increase the encapsulation efficiency. The comparatively higher encapsulation efficiency is due to the high bond affinity and hydrophilicity of nano-HA particles. When hydrophilic drugs such as gentamicin are distributed over the HA phase the amount of drug moving toward the aqueous phase is reduced. Summarising, it is like increasing the system affinity by the drugs, providing higher encapsulation efficiency. The control on the release rate is also enhanced, since the interaction of gentamicin with nano-HA avoids the initial burst effect and allows a sustained release for more than 10 weeks.

Calcium-deficient hydroxyapatites nanoparticles (nano-CDHA) have been also used to regulate the kinetic release of chitosan-derived microspheres.[234] Carboxymethyl-hexanoyl chitosan constitutes a hydrophilic matrix with a burst-release profile in a highly swollen state. Incorporation of nano-CDHA has demonstrated the ability to regulate the release of ibuprofen as the nanoparticles amount increases in the composite, due to the inorganic nanofiller acting as a crosslink agent and diffusion barrier. On the contrary, when nano-CDHA is incorporated with O-hexanoyl chitosan matrix, which is a hydrophobic compound, the ibuprofen release is accelerated. This fact can be explained in terms of higher polymer degradation due to the hydrophilic character of nano-CDHA, facilitating the water accessibility and thus enhancing the drug diffusion. The amount of nano-CDHA incorporated into chitosan matrices is not the only factor that alters the extent of filler–polymer interactions. The drug-release behaviour of nano-CDHA/chitosan is strongly dependent on the synthesis method.[235] For instance, the diffusion exponent of the CDHA/chitosan membranes is lower for that synthesised through the *ex-situ* processes, *i.e.* for those nanocomposites where CDHA nanofiller was synthesised first and then added into the chitosan solution. When these nanocomposites are prepared as membranes, the permeability is lower when the nano-CDHA is synthesised in the presence of chitosan, which can be explained in terms of a better dispersion of the CDHA nanoparticles, resulting in a more efficient physical barrier.

The use of silver has recently become one of the preferred methods to impede microbial proliferation on biomaterials and medical devices. Silver and silver-based compounds are highly antimicrobial to as many as 16 kinds of bacteria, including *Escherichia coli* and *Staphylococcus aureus*.[236] Silver-loaded HA powder has shown antibacterial effects, both in nutrient-rich and poor environments.[237] From a crystal-chemical point of view, the substitution of Ag^+ (1.28 Å) ions takes place for Ca^{2+} (0.99 Å) preferentially in the Ca(1) site of HA, and this leads to an increase in the lattice parameters linearly with the amount of silver added in the range of atomic ratio Ag/(Ag + Ca) between 0 and 0.055.[238] The proposed general formula is $Ca_{10-x}Ag_x(PO_4)_6(OH)_{2-x}\square_x$, with vacancies at the hydroxyl site due to charge imbalance caused by Ag^+ for

Ca^{2+} ions, although PO_4^{3-} for HPO_4^{2-} is also a likely substitution to compensate the charge imbalance. Rameshbabu *et al.*[239] have synthesised nanosized silver-substituted HA (AgHA) using microwaves and incorporating $Ag(NO)_3$ to the reaction media in the required stoichiometric amounts. The microwave synthesis is a fast, simple, and efficient method to prepare nanosized materials,[240] with narrow particle-size distribution due to fast homogeneous nucleation.[241] Ag-substituted nano-HA prepared with this method are nanosized with needle-like morphology, with width ranging from 15 to 20 nm and length around 60–70 nm. A substitution degree of $x = 0.05$ in the general formula $Ca_{10-x}Ag_x(PO_4)_6(OH)_{2-x}\square_x$ is enough to completely inhibit the growth of *E. coli* and *S. aureus* after 24 and 48 h with 10^5 cells/mL, while exhibiting excellent osteoblast adherence and spreading.

4.5.2.3 Nanoapatites for Nonviral Gene-Delivery Systems

Gene therapy is becoming a rapidly growing therapeutic strategy, which consists of *transfecting* a modified gene into the genome to replace a disease-causing gene.[242,243] The incorporation of bare DNA would be the simplest method of transfection. However, in these cases the gene expression is very low due to the low endosomal escape, nuclease degradation and inefficient nuclear uptake.[244,245] Consequently, other methods must be applied and for these purposes a *vector* must be used to deliver the therapeutic gene to the patient's target cells. Efficient gene transfection is achieved when the gene-delivery vector facilitates:

- Physical and chemical stability to the DNA in the extracellular space;
- Cellular uptake;
- DNA escape from the endosomal network;
- Cytosolic transport;
- Nuclear localisation of the DNA for transcription.[246,247]

Currently, the most common vectors are viruses that have been genetically altered to carry normal human DNA. These vectors infect the cell of the patient and unload the genetic material containing the therapeutic human gene into the target cell, restoring it to a normal state.

The development of nonviral vectors is currently catching the attention of many researchers. Nonviral methods present certain advantages over viral ones, like simple large-scale production and low host immunogenicity among others. The major limitations of nonviral gene transfer lie in that it must be tailored to overcome the intracellular barriers to DNA delivery, including the cellular and nuclear membranes.[248] However, recent advances in vector technology have yielded molecules and techniques with transfection efficiencies similar to those of viruses. Among these techniques, the binding of DNA to calcium phosphates is one of the most attractive options, due to the high biocompatibility, biodegradability, easy handling and high adsorptive capacity for plasmid DNA (pDNA).[249–252]

The hydrothermal technique has been applied to prepare hydroxyapatite nanoparticles to evaluate as a material for pDNA transfection.[253] The pDNA binding to a previously obtained nano-HA can be achieved through the mixture of a nanoparticles suspension with the pDNA and subsequent incubation at room temperature. The DNA–nanoparticle complexes transfect pDNA into SGC-7901 mice cells *in vitro* and TEM examination demonstrated their bio-distribution and expression within the cytoplasm and also a little in the nuclei of the liver, kidney and brain tissue cells of mice. The ratio potential to adsorb pDNA is about 1:36 and only can be carried out under acidic and neutral conditions.

pDNA encapsulated in calcium phosphate nanoparticles has been prepared as a DNA delivery carrier which has specifically targeted these particles to liver cells after appropriate surface modification.[254,255] Although the pDNA en-trapped in these nanoparticles is highly protected from enzymatic degradation, the transfection efficiency of these synthetic systems is not optimal. The problem lies in the fact that prolonged ultrasonication used to be a prerequisite for redispersion of nanoparticles in aqueous buffers, which led to the partial disintegration of DNA molecules, thus reducing the transfection efficiency. In order to overcome this situation, these methods have been optimised by forming the calcium phosphate within the droplets of microemulsions and completely avoiding ultrasonication.[256]

Olton *et al.*[257] have dealt with the influence of synthesis parameters on transfection efficiency. The results of these authors revealed that improved, more consistent levels of gene expression can be achieved by optimising both the stoichiometry (Ca/P ratio) of the CaP particles as well as the mode in which the precursor solutions are mixed. The optimised forms of these CaP particles were approximately 25–50 nm in size (when complexed with pDNA) and were effi-cient at both binding and condensing the genetic material. Differences in gene expression are not only due to a change in size of the naked CaP particles but are rather due to the combined effects of pDNA binding and condensation to the particle, which ultimately dictates the overall size of the pDNA–NanoCaPs complex size.

References

1. T. Traykova, C. Aparicio, M. P. Ginebra and J. A. Planell, *Nanomedicine*, 2006, **1**, 91.
2. D. Aronov, R. Rosen, E. Z. Ron and G. Rosenman, *Proc. Biochem.*, 2006, **41**, 2367.
3. K. Kawanabe, H. Iida, Y. Matsusue, H. Nishimatsu, R. Kasai and T. Nakamura, *Acta Orthop Scand.*, 1998, **69**, 237.
4. B. Ben-Nissan, *MRS Bull*, 2004, **29**, 28.
5. Y. F. Chou, W. Huang, J. C. Y. Duna, T. A. Millar and B. M. Wu, *Biomaterials*, 2005, **26**, 285.
6. S. Bose and S. K. Saha, *Chem. Mater.*, 2003, **15**, 4464.

7. EFFO and NOF (1997). *Osteoporos Int.* 1997, 7, 1.
8. A. Randell, P. N. Sambrook and T. V. Nguyen, *Osteoporos Int.*, 1995, **5**, 427.
9. L. M. Rodríguez-Lorenzo and M. Vallet-Regí, *Chem. Mater.*, 2000, **12**, 2460.
10. A. Toni, C. G. Lewis and A. Sudanese, *J. Arthroplasty*, 1994, **9**, 435.
11. L. L. Hench and J. Wilson, In: *Bioceramics*, R. Z. LeGeros, J. P. LeGeros ed., vol. 11. World Scientific, New York City, NY,1998. p. 31.
12. R. D. Bloebaum and J. A. Dupont, *J. Arthroplasty*, 1993, **8**, 195.
13. A. R. Biesbrock and M. Edgerton, *Int. J. Oral Maxillofac Implants*, 1995, **10**, 712.
14. E. W. Morscher, A. Hefti and U. Aebi, *J. Bone Jt. Surg. Br.*, 1998, **80**, 267.
15. T. Ichikawa, K. Hirota and H. Kanitani, *J. Oral Implantol.*, 1996, **22**, 232.
16. T. J. Webster and J. U. Ejiofor, *Biomaterials*, 2004, **25**, 4731.
17. T. J. Webster, C. Ergun, R. H. Doremus, R. W. Siegel and R. Bizios, *Biomaterials*, 2000, **21**, 1803.
18. X. Wang, X. Shen, X. Li and C. M. Agarwal, *Bone*, 2002, **31**, 1.
19. T. J. Webster, R. W. Siegel and R. Bizios, *Nanostruct. Mater.*, 1999, **12**, 983.
20. T. J. Webster, R. W. Siegel and R. Bizios, *Soc. Biomater. Trans.*, 1999, **1**, 88.
21. T. J. Webster, C. Ergun, R. H. Doremus, R. W. Siegel and R. Bizios, *J. Biomed. Mater. Res.*, 2000, **51**, 475.
22. S. Ayad S, R. Boot–Handford, M. J. Humpries, K. E. Kadler and A. Shuttleworth, *The Extracellular Matrix Facts Book*, Academic Press Inc., San Diego, 1994, p. 29.
23. Y. R. Cai, Y. K. Liu, W. Q. Yan, Q. H. Hu, J. H. Tao, M. Zhang, Z. L. Shi and R. K. Tang, *J. Mater. Chem.*, 2007, **17**, 3780.
24. A. Scabbia, *J. Clin. Periodontol.*, 2004, **31**, 348.
25. W. Sun, C. Chu, J. Wuang and H. Zhao, *J. Mater. Sci.: Mater. Med.*, 2007, **18**, 677.
26. M. Vallet-Regí and J. M. Gónzalez-Calbet, *Prog. Solid State Chem.*, 2004, **32**, 1.
27. M. Vallet-Regí, *J. Chem. Soc. Dalton Trans.*, 2001, **2**, 97.
28. S. V. Dorozhkin and M. Epple, *Angew. Chem. Int. Ed.*, 2002, **41**, 3130.
29. M. Yoshimura, H. Suda, K. Okamoto and K. Ioku, *J. Mater. Sci.*, 1994, **29**, 3399.
30. S. Mann, *Biomineralization: Principles and Concepts in Bioinorganic Materials Chemistry*, Oxford University Press, Oxford, U.K., 2001.
31. E. M. Burke, Y. Guo, L. Colon, M. Rahima, A. Veis and G. H. Nancollas, *Colloids Surf., B*, 2000, **17**, 49.
32. A. Bigi, E. Boanini, D. Walsh and S. Mann, *Angew. Chem., Int. Ed.*, 2002, **41**, 2163.
33. E. Bettoni, A. Bigi, G. Falini, S. Panzavolta and N. Roveri, *J. Mater. Chem.*, 1999, **9**, 779.

34. D. Walsh, J. L. Kingston, B. R. Heywood and S. Mann, *J. Cryst. Growth*, 1993, **133**, 1.
35. S. Sarda, M. Heughebaert and A. Lebugle, *Chem. Mater.*, 1999, **11**, 2722.
36. M. Bujan, M. Sikiric, F. Vincekovic, N. Vdovic, N. Garti and F. H. Milhofer, *Langmuir*, 2001, **17**, 6461.
37. L. Hovarth, I. Smit, M. Sikiric and F. Vincekovic, *J. Cryst. Growth*, 2000, **219**, 91.
38. L. Qi, J. Ma, H. Cheng and Z. Zhao, *J. Mater. Sci. Lett.*, 1997, **16**, 1779.
39. M. Antonietti, M. Breulmann, C. Goltner, H. Colfen, K. Wong, D. Walsh and S. Mann, *Chem.-Eur. J.*, 1998, **4**, 2493.
40. H. A. Schmidt and A. E. Ostafin, *Adv. Mater.*, 2002, **14**, 532.
41. W. Zhang, S. S. Liao and F. Z. Cui, *Chem. Mater.*, 2003, **15**, 3221.
42. S. Mann, D. D. Archibald, J. M. Didymus, T. Douglass, B. R. Heywood and F. C. Meldrum, *Science*, 1993, **261**, 1286.
43. S. Mann, *Nature*, 1993, **365**, 499.
44. D. D. Archibald and S. Mann, *Nature*, 1993, **364**, 430.
45. I. Weissbuch, F. Frolow, L. Addadi, M. Lahav and L. Leiserowitz, *J. Am. Chem. Soc.*, 1990, **112**, 7718.
46. A. Firouzi, D. Kumar, L. M. Bull, T. Besier, P. Sieger and Q. Huo, *Science*, 1995, **267**, 1138.
47. X. Wang, J. Zhuang, Q. Peng and Y. Li, *Adv. Mater.*, 2006, **18**, 2031.
48. Y. Wang, S. Zhang, K. Wei, N. Zhao, J. Chen and X. Wang, *Mater. Lett.*, 2006, **60**, 1484.
49. F. C. Meldrum, N. A. Kotov and J. H. Fendler, *J. Phys. Chem.*, 1994, **98**, 4506.
50. D. H. Gray, S. Hu, E. Juang and D. L. Gin, *Adv. Mater.*, 1997, **9**, 731.
51. L. Yan, Y. D. Li, Z. X. Deng, J. Zhuang and X. M. Sun, *Int. J. Inorg. Mater.*, 2001, **3**, 633.
52. K. Onuma and A. Ito, *Chem. Mater.*, 1998, **10**, 3346.
53. J. D. Chen, Y. J. Wang, K. Wei, S. H. Zhang and W. T. Shi, *Biomaterials*, 2007, **28**, 2275.
54. K. S. Katti, *Col. Surf. B*, 2004, **39**, 143.
55. L. G. Gutwein and T. J. Webster, *Biomaterials*, 2004, **25**, 4175.
56. B. H. Kear, J. Coliazzi and W. E. Mayo, *Scr. Mater.*, 2001, **44**, 2065.
57. C. Piconi and G. Maccauro, *Biomaterials*, 1999, **20**, 1.
58. Y. M. Sung, Y. K. Shin and J. J. Ryu, *Nanotechnology*, 2007, **18**, 065602.
59. W. Suchanek and M. Yoshimura, *J. Mater. Res.*, 1998, **13**, 94.
60. T. J. Webster, R. W. Siegel and R. Bizios, *Biomaterials*, 1999, **20**, 1222.
61. I. B. Lonor, A. Ito, K. Onuma, N. Kanzaki and R. L. Reis, *Biomaterials*, 2003, **24**, 579.
62. A. R. Boccaccini and V. Maquet, *Compos. Sci. Tech.*, 2003, **63**, 2417.
63. U. Arnold, K. Lindenhayn and C. Perka, *Biomaterials*, 2002, **23**, 2303.
64. N. Tamai, A. Myoui, M. Hirao, T. Kaito, T. Ochi, J. Tanaka, K. Takaoka and H. Yoshikawa, *Osteoarthr Cartilage*, 2005, **13**, 405.
65. C. Doyle, E. T. Tanner and W. Bonfield, *Biomaterials*, 1991, **12**, 841.
66. J. Ni and M. Wang, *Mater. Sci. Eng. C.*, 2002, **20**, 101.

67. Y. E. Greish, J. D. Bender, S. Lakshmi, P. W. Brown, H. R. Allcock and C. T. Laurencin, *Biomaterials*, 2005, **26**, 1.
68. D. Choi, K. G. Marra and P. N. Kumta, *Mater. Res. Bull.*, 2004, **39**, 417.
69. R. A. Sousa, R. L. Reis, A. M. Cunha and M. J. Bevis, *Compos. Sci. Technol.*, 2003, **63**, 389.
70. A. P. Marques and R. L. Reis, *Mater. Sci. Eng. C.*, 2005, **25**, 215.
71. X. M. Deng, J. Y. Hao and C. S. Wang, *Biomaterials*, 2001, **22**, 2867.
72. Z. K. Hong, P. B. Zhang, C. L. He, X. Y. Qiu, A. X. Liu, L. Chen, X. S. Chen and X. B. Jing, *Biomaterials*, 2005, **26**, 6296.
73. J. H. Lee, T. G. Park, H. S. Park, D. S. Lee, Y. K. Lee, S. C. Yoon and J. D. Nam, *Biomaterials*, 2003, **24**, 2773.
74. S. S. Kim, M. S. Park, Q. Jeon, C. Y. Choi and B. S. Kim, *Biomaterials.*, 2006, **27**, 1399.
75. Q. L. Hu, B. Q. Li, M. Wang and J. C. Shen, *Biomaterials.*, 2004, **25**, 779.
76. C. Du, F. Z. Cui, Q. L. Feng, X. D. Zhu and K. de Groot, *J. Biomed. Mater. Res.*, 1998, **42**, 540.
77. M. Kikuchi, S. Itoh, S. Ichinose, K. Shinomiya and J. Tanaka, *Biomaterials*, 2001, **22**, 1705.
78. M. Kikuchi, H. N. Matsumoto, T. Yamada, Y. Koyama, K. Takakuda and J. Tanaka, *Biomaterials*, 2004, **25**, 63.
79. A. K. Lynn, T. Nakamura, N. Patel, A. E. Porter, A. C. Renouf, P. R. Laity, S. M. Best, R. E. Cameron, Y. Shimizu and W. Bonfield, *J. Biomed. Mater. Res.*, 2005, **74A**, 447.
80. M. C. Chang and J. Tanaka, *Biomaterials*, 2002, **23**, 4811.
81. M. C. Chang and J. Tanaka, *Biomaterials*, 2002, **23**, 3879.
82. S. S. Liao, F. Z. Cui and Y. Zhu, *J. Bioact. Compat. Polym.*, 2004, **19**, 117.
83. M. C. Chang, C. C. Ko and W. H. Douglas, *Biomaterials*, 2003, **24**, 2853.
84. M. C. Chang, C. C. Ko and W. H. Douglas, *Biomaterials*, 2003, **24**, 3087.
85. Kim Hae-Won, Kim Hyoun-Ee and Veid Salih, *Biomaterials*, 2005, **26**, 5221.
86. J. Y. Hao, L. Yu, Z. Shaobing, L. Zhen and D. Xianmo, *Biomaterials*, 2003, **24**, 1531.
87. T. A. Taton, *Nature*, 2001, **412**, 491.
88. J. D. Hartgerink, E. Beniash and S. I. Stupp, *Science*, 2001, **294**, 1684.
89. G. N. Ramachandran and A. H. Reddi, *Biochemistry of Collagen*, Plenum Press, New York, 1976.
90. N. Degirmenbasi, D. M. Kalyon and E. Birinci, *Colloid Surface B.*, 2006, **48**, 42.
91. N. A. Peppas and N. K. Mongia, *Eur. J. Pharm. Biopharm.*, 1997, **43**, 51.
92. T. H. Young, W. Y. Chuang, M. Y. Hsieh, L. W. Chen and J. P. Hsu, *Biomaterials*, 2002, **23**, 3495.
93. M. Kobayashi, J. Toguchida and M. Oka, *Biomaterials*, 2003, **24**, 639.
94. H. Jiang, G. Campbell, D. Boughner, W. K. Wan and M. Quantz, *Med. Eng. Phys.*, 2004, **26**, 269.
95. H. Jiang, G. Campbell and F. Xi, *Med. Eng. Phys.*, 2005, **27**, 175.

96. H. L. Nichols, N. Zhang, J. Zhang, D. Shi, S. Bhaduri and X. Wen, *J. Biomed. Mater. Res.*, 2007, **82A**, 373.
97. L. O'Toole, A. J. Beck, A. P. Ameen, F. R. Jones and R. D. Short, *J. Chem. Soc. Faraday Trans.*, 1995, **91**, 3907.
98. M. R. Alexander and T. M. Duc, *J. Mater Chem*, 1998, **8**, 937.
99. A. J. McManus, R. H. Doremus, R. W. Siegel and R. Bizios, *J. Biomed. Mater. Res.*, 2005, **72A**, 98.
100. T. M. Keaveny and W. C. Hayes, *Bone*, 1993, **7**, 285.
101. C. L. Wu, W. G. Weng, D. J. Wu and W. L. Yan, *Polymer*, 2003, **44**, 1781.
102. A. G. Rozhin, Y. Sakakibara, M. Tokumoto, H. Kataura and Y. Achiba, *Thin Solid Films*, 2004, **464–465**, 368.
103. D. Z. Chen, C. Y. Tang, K. C. Chan, C. P. Tsui, P. H. F. Yu, M. C. P. Leung and P. S. Uskokovic, *Compos. Sci. Techol.*, 2007, **67**, 1617.
104. H. W. Kim, *J. Biomed. Mater. Res.*, 2007, **83A**, 169.
105. T. J. Webster, R. W. Siegel and R. Bizios, In *Bioceramics 11.*, R. Z. LeGeros, J. P. LeGeros ed., World Scientific Publishing Co., New York, USA, 1998, 273–276.
106. J. B. Brunski, *Clin Mater*, 1992, **10**, 153.
107. G. Heimke, *Osseo-integrated Implants Volume I: Basics, Materials and Joint Replacements.*, CRC Press, Boca Raton, FL, 1990, p. 31–80.
108. P. Griss, M. H. Hackenbroch, M. Jager, B. Preussner, T. Schafer, R. Seebauer, W. van Eimeren and W. Winkler, *Aktuelle Probl Chir Orthop.*, 1982, **21**, 1.
109. T. J. Webster, E. A. Massa-Schlueter, J. L. Smith and E. B. Slamovich, *Biomaterials*, 2004, **25**, 2111.
110. J. P. Bearinger, D. G. Castner and K. E. Healy, *J. Biomater. Sci. Polym. Ed.*, 1998, **7**, 652.
111. K. C. Dee, T. T. Andersen and R. Bizios, *J. Biomed. Mater. Res.*, 1998, **40**, 371.
112. E. R. Kenawy, G. L. Bowlin, K. Mansfield, J. Layman, D. G. Simpson, E. H. Sanders and G. E. Wnek, *J. Control. Rel.*, 2002, **81**, 57.
113. B. M. Min, L. Jeong, Y. S. Nam, J. M. Kim and W. H. Park, *Intl. J. Biol. Macromol.*, 2004, **34**, 281.
114. H. Yoshimoto, Y. M. Shin, H. Terai and J. P. Vacanti, *Biomaterials*, 2003, **24**, 2077.
115. W. J. Li, R. Tuli, C. Okafor, A. Derfoul, K. G. Danielson, D. J. Hall and R. S. Tuan, *Biomaterials*, 2005, **26**, 599.
116. A. Formhals, Process and apparatus for preparing artificial threads, US Patent 1975504, 1934.
117. P. Wutticharoenmongkol, N. Sanchavanikit, P. Pavasant and P. Supaphol, *Macromol. Biosci.*, 2006, **6**, 70.
118. P. Wutticharoenmongkol, P. Pavasant and P. Supaphol, *Biomacromolecules*, 2007, **8**, 2602.
119. H. W. Kim, H. H. Lee and J. C. Knowles, *J. Biomed. Mater. Res.*, 2006, **79A**, 643.

120. G. Sui, X. Yang, F. Mei, X. Hu, G. Chen, X. Deng and S. Ryu, *J. Biomed. Mater. Res.*, 2007, **82A**, 445.
121. J. K. Vasir, K. Tambwekar and S. Garg, *Int. J. Pharm.*, 2003, **255**, 13.
122. U. Edlund and A. C. Albertsson, *Adv. Polym. Sci.*, 2002, **157**, 67.
123. H. Kawaguchi, *Prog. Polym. Sci.*, 2000, **25**, 1171.
124. S. Freiberg and X. X. Zhu, *Int. J. Pharm.*, 2004, **282**, 1.
125. E. Ruiz-Hernández, A. López-Noriega, D. Arcos, I. Izquierdo-Barba, O. Terasaki and M. Vallet-Regí, *Chem. Mater.*, 2007, **19**, 3455.
126. X. Y. Qiu, Z. K. Hong, J. L. Hu, L. Chen, X. S. Chen and X. B. Ping, *Biomacromolecules*, 2005, **6**, 1193.
127. X. Qiu, Y. Han, X. Zhuang, X. Chen, Y. Li and X. Jing, *J. Nanoparticle Res.*, 2007, **9**, 901.
128. D. Arcos, C. V. Ragel and M. Vallet-Regí, *Biomaterials*, 2001, **22**, 701.
129. D. Arcos, J. Peña and M. Vallet-Regí, *Chem. Mater.*, 2003, **15**, 4132.
130. C. V. Ragel and M. Vallet-Regí, *J. Biomed. Mater. Res.*, 2000, **51**, 424.
131. A. Rámila, R. P. del Real, R. Marcos, P. Horcajada and M. Vallet-Regí, *J. Sol-Gel Sci Technol.*, 2003, **26**, 1195.
132. S. Padilla, R. P. del Real and M. Vallet-Regí, *J. Control. Rel.*, 2002, **83**, 343.
133. S. N. Khorasani, S. Deb, J. C. Behiri, M. Braden and W. Bonfield, *Bioceramics*, 1992, **5**, 225.
134. Y. M. Khan, D. S. Katti and C. T. Laurencin, *J. Biomed. Mater. Res.*, 2004, **69A**, 728.
135. A. Piattelli, M. Franco, G. Ferronato, M. T. Santello, R. Martinetti and A. Scarano, *Biomaterials*, 1997, **18**, 629.
136. S. A. Brown, L. Farnsworth, K. Merrit and T. D. Crowe, *J. Biomed. Mater. Res.*, 1988, **22**, 321.
137. L. L. Hench and J. Wilson, *Biomater. Sci.*, 1984, **226**, 630.
138. W. Suchanec and M. Yoshimura, *J. Mater. Res.*, 1998, **13**, 94.
139. R. Z. LeGeros, *Clin. Orthop. Relat. Res.*, 2002, **395**, 81.
140. L. Sun, C. C. Berndt, K. A. Gross and A. Kucuk, *J. Biomed. Mater. Res: Appl. Biomat*, 2001, **58**, 570.
141. Y. Yang, K. H. Kim and J. L. Ong, *Biomaterials*, 2005, **26**, 327.
142. F. J. García-Sanz, M. B. Mayor, J. L. Arias, J. Pou, B. León and M. Pérez-Amor, *J. Mater. Sci.: Mater. Med.*, 1997, **8**, 861.
143. J. G. C. Wolke, J. P. C. M. van der Waerden, H. G. Schaeken and J. A. Jansen, *Biomaterials*, 2003, **24**, 2623.
144. Z. S. Luo, F. Z. Cui and W. Z. Li, *J. Biomed. Mater. Res.*, 1999, **46**, 80.
145. H. Ishizawa and M. Ogino, *J. Biomed. Mater. Res.*, 1997, **34**, 15.
146. R. Chiesa, E. Sandrini, M. Santin, G. Rondelli and A. Cigada, *J. Appl. Biomater. Biomech.*, 2003, **1**, 91.
147. A. Bigi, B. Bracci, F. Cuisinier, R. Elkaim, R. Giardino, I. Mayer, I. N. Mihailescu, G. Socol L. Sturba and P. Torricelli, *Biomaterials*, 2005, **26**, 2381.
148. M. V. Cabañas and M. Vallet-Regi, *J. Mater. Chem.*, 2003, **13**, 1104.

149. M. Cifuentes, M. V. Cabañas and M. Vallet-Regí, *Key. Eng. Mater.*, 2001, **192–195**, 135.

150. L. L. Hench and J. K. West, *Chem. Rev.*, 1990, **90**, 33.

151. C. J. Brinker and G. W. Scherer, *Sol-Gel Science: The Physics and Chemistry of Sol-Gel Processing.*, Academic Press, San Diego, 1990.

152. N. Hijón, M. V. Cabañas, I. Izquierdo-Barba and M. Vallet-Regí, *Chem. Mater.*, 2004, **16**, 1451.

153. T. Brendel, A. Engel and C. Rüssel, *J. Mater. Sci.: Mater. Med.*, 1992, **3**, 175.

154. S. W. Russell, K. A. Luptak, C. T. Suchicital, T. L. Alford and V. C. Pizziconi, *J. Am. Ceram. Soc.*, 1996, **79**, 837.

155. M. Hsieh, L. Perng and T. Chin, *Mater. Chem. Phys.*, 2002, **74**, 245.

156. L. Goins, S. Holliday, A. Staniskevsky, *Mater. Res. Soc. Symp. Proc.*, 2004, vol. EXS-1, H6.301-3.

157. C. You and S. Kim, *J. Sol-Gel Sci. Technol.*, 2001, **21**, 49.

158. K. Hwang and Y. Lim, *Surf. Coat. Technol.*, 1999, **115**, 172.

159. Y. Kojima, A. Shiraishi, K. Ishii, T. Yasue and Y. Arai, *Phosphorus Res. Bull.*, 1993, **3**, 79.

160. D. Liu, T. Troczynski and W. J. Tseng, *Biomaterials*, 2001, **22**, 1721.

161. L. Gan and R. Pilliar, *Biomaterials*, 2004, **25**, 5303.

162. L. D. Piveteau, M. I. Girona, L. Schlapbach, P. Barboux, J. P. Boilot and B. Gasser, *J. Mater. Sci.: Mater. Med.*, 1999, **10**, 161.

163. M. Cavalli, G. Gnappi, A. Montenero, D. Bersani, P. P. Lottici, S. Kaciulis, G. Mattogno and M. Fini, *J. Mater. Sci.*, 2001, **36**, 3253.

164. W. Weng and J. L. Baptista, *Biomaterials*, 1998, **19**, 125.

165. C. S. Chai, K. A. Gross and B. Ben-Nissan, *Biomaterials*, 1998, **19**, 2291.

166. K. A. Gross, C. S. Chai, G. S. K. Kannangara, B. Ben-Nissan and L. Hanley, *J. Mater. Sci.: Mater. Med.*, 1998, **9**, 839.

167. D. B. Haddow, P. F. James and R. Van Noort, *J. Sol-Gel Sci. Technol.*, 1998, **13**, 261.

168. B. Ben-Nissan, A. Milev and R. Vago, *Biomaterials*, 2004, **25**, 4971.

169. E. Tkalcec, M. Sauer, R. Nonninger and H. Schmidt, *J. Mater. Sci.*, 2001, **36**, 5253.

170. I. Izquierdo-Barba, N. Hijón, M. V. Cabañas and M. Vallet-Regí, *Key Eng. Mater.*, 2004, **254–256**, 363.

171. N. Hijón, M. V. Cabañas, I. Izquierdo-Barba, M. A. García and M. Vallet-Regí, *Solid State Sci.*, 2006, **8**, 685.

172. S. J. Lin, R. Z. LeGeros and J. P. LeGeros, *J. Biomed. Mater. Res.*, 2003, **66A**, 819.

173. Y. W. Gu, K. A. Khor and P. Cheang, *Biomaterials*, 2003, **24**, 1603.

174. R. Z. LeGeros, *Calcium Phosphates in Enamel, Dentin and Bone*, in: H. M. Myers ed., *Calcium Phosphates in Oral Biology in Medicine*, in: *Monographs in Oral Science*, Karge, Zurich, 1991, pp. 108–129.

175. M. Vallet-Regí, C. V. Ragel and A. J. Salinas, *Eur. J. Inorg. Chem.*, 2003, 1029.

176. N. Hijón, M. V. Cabañas, J. Peña and M. Vallet-Regí, *Acta Biomater.*, 2006, **2**, 567.
177. I. Ichinose, H. Senzu and T. Kunitake, *Chem. Lett.*, 1996, **257**, 258.
178. J. He, I. Ichinose, S. Fujikawa, T. Kunitake and A. Nakao, *Chem Mater*, 2002, **14**, 3493.
179. K. Acharya and T. Kunitake, *Langmuir*, 2003, **19**, 2260.
180. P. Li, *J. Biomed. Mater. Res.*, 2003, **66A**, 79.
181. J. D. De Bruijn and C. A. Van Blitterswijk, In *Biomaterials in Surgery.*, G. Walenkamp ed., Geory Thieme Verlag, Stuttgart, 1998, pp. 77–72.
182. C. M. Agrawal, J. Best, J. D. Heckman and B. D. Boyan, *Biomaterials*, 1995, **16**, 1255.
183. Y. Liu, P. Layrolle, J. de Bruijn, C. van Blitterswijk and K. De Groot, *J. Biomed. Mater. Res.*, 2001, **57**, 327.
184. S. Leeuwenburgh, P. Layrolle, F. Barrere, J. de Bruijn, J. Schoonman, C. A. van Blitterswijk and K. de Groot, *J. Biomed. Mater. Res.*, 2001, **56**, 208.
185. F. Barrère, C. M. van der Valk, R. A. J. Dalmeijer, G. Meijer, C. A. van Blitterswijk, K. De Groot and P. Layrolle, *J. Biomed. Mater. Res.*, 2003, **66A**, 779.
186. K. K. W. Lo, T. K. M. Lee, J. S. Y. Lau, W. L. Poon and S. H. Cheng, *Inorg. Chem.*, 2008, **47**, 200.
187. K. Hanaoka, K. Kikuchi, S. Kobayashi and T. Nagano, *J. Am. Chem. Soc.*, 2007, **129**, 13502.
188. K. K. V. Lo, W. K. Hui, C. K. Chung, K. H. K. Tsang, D. C. M. Ng, N. Y. Zhu and K. K. Cheung, *Coord. Chem. Rev.*, 2005, **249**, 1434.
189. H. M. E. Azzazy, M. M. H. Manssur and S. C. Kazmierczak, *Clin Biochem.*, 2007, **40**, 917.
190. A. P. Alivisatos, W. Gu and C. Larabell, *Annu Rev Biomed Eng*, 2005, **7**, 55.
191. D. E. Clapham, *Cell*, 1995, **80**, 259.
192. Y. Kakizawa, S. Furukawa and K. Kataoka, *J. Control. Rel.*, 2004, **97**, 345.
193. A. Doat, M. Fanjul, F. Pellé, E. Hollande and A. Lebugle, *Biomaterials*, 2003, **24**, 3365.
194. A. Doat, F. Pellé, N. Gardant and A. Lebugle, *J. Solid State Chem.*, 2004, **177**, 1179.
195. A. Lebugle, F. Pellé, C. Charvillat, I. Rousselot and J. Y. Chane-Ching, *Chem. Commun.*, 2006, **606**.
196. V. P. Torchilin, *Nature Rev.*, 2005, **4**, 145.
197. J. W. Yoo and C. H. Lee, *J. Control. Rel.*, 2006, **112**, 1.
198. M. Malmeten, *Soft Mater.*, 2006, **2**, 760.
199. M. Vallet-Regí, *Chem. Eur. J.*, 2006, **12**, 5934.
200. M. Vallet-Regí, F. Balas and D. Arcos, *Angew. Chem. Int. Ed.*, 2007, **46**, 7548.
201. M. Vallet-Regí, A. Rámila, R. P. del Real and J. Pérez-Pariente, *Chem. Mater.*, 2001, **13**, 308.

202. F. Balas, M. Manzano, P. Horcajada and M. Vallet-Regí, *J. Am. Chem. Soc.*, 2006, **128**, 8116.

203. M. Vallet-Regí, *Dalton Trans.*, 2006, **1**, 5211.

204. B. Muñoz, A. Rámila, J. Pérez-Pariente, I. Díaz and M. Vallet-Regí, *Chem. Mater.*, 2003, **15**, 500.

205. F. Lamoureux, V. Trichet, C. Chipoy, F. Blanchard, F. Gouin and F. Redini, *Expert Rev. Anticancer Ther.*, 2007, **7**, 169.

206. P. K. Bajpai and H. A. Benghuzzi, *J. Biomed. Mater. Res.*, 1988, **22**, 1245.

207. E. P. Goldberg, A. R. Hadba, B. A. Almond and J. S. Marotta, *J. Pharm. Pharmacol.*, 2002, **54**, 159.

208. K. J. Harrington, F. Rowlinson-Busza and K. N. Syringos, *Clin. Cancer. Res.*, 2000, **6**, 2528.

209. A. Lebugle, A. Rodrigues, P. Bonnevialle, J. J. Voigt, P. Canal and F. Rodriguez, *Biomaterials*, 2002, **23**, 3517.

210. K. O. Lillehei. Q. Kong, S. J. Withrow and B. Kleinschmidt-DeMasters, *Neurosurgery*, 1996, 1191.

211. S. Miura, Y. Mii and Y. Miyauchi, *Jpn. J. Clin. Oncol.*, 1995, **25**, 61.

212. R. C. Straw, S. J. Withrow and E. B. Douple, *J. Orthop. Res.*, 1994, **12**, 1.

213. Y. Tahara and Y. Ishii, *J. Orthop. Sci.*, 2001, **6**, 556.

214. A. Uchida, Y. Shinto, N. Araki and K. Ono, *J. Orthop. Res.*, 1992, **10**, 440.

215. A. Barroug, L. T. Kuhn, L. C. Gerstenfeld and M. J. Glimcher, *J. Orthop. Res.*, 2004, **22**, 703.

216. A. Barroug and M. J. Glimcher, *J. Orthop. Res.*, 2002, **20**, 274.

217. A. Barroug, J. Lemaitre and P. G. Rouxhet, *Colloids. Surf.*, 1989, **37**, 339.

218. A. Barroug, E. Lernous, J. Lemaitre and P. G. Rouxhet, *J. Colloid. Interf. Sci.*, 1998, **208**, 147.

219. J. Guicheux, G. Grimandi and M. Trecant, *J. Biomed. Mater. Res.*, 1997, **34**, 165.

220. V. C. Honnorat-Benabbou, A. Lebugle, B. Sallek and D. Lagarrigue, *J. Mater. Sci.*, 2001, **12**, 107.

221. B. Palazzo, M. Iafisco, M. Laforgia, N. Margiotta, G. Natile, C. L. Bianchi, D. Walsh, S. Mann and N. Roveri, *Adv. Funct. Mater.*, 2007, **17**, 2180.

222. S. P. A. Guaber, G. Gazzaniga, N. Roveri, L. Rimondini, B. Palazzo, M. Iafisco, P. Gualandi, *EU Patent* 005 146, 2006.

223. E. Landi, A. Tampieri, G. Celotti and S. Sprio, *J. Eur. Ceram. Soc.*, 2000, **20**, 2377.

224. L. Sz-Chian, C. San-Yuan, L. HsinYi and B. Jong-Shing, *Biomaterials*, 2004, **25**, 189.

225. F. Wingen and D. Schmahl, *Drug Res.*, 1985, **35**, 1565.

226. M. J. Bloemink, B. K. Keppler, H. Zahn, J. P. Dorenbos, R. J. Heetebrij and J. Reedijk, *Inorg. Chem.*, 1994, **33**, 1127.

227. T. Klenner, P. Valenzuela-Paz, B. K. Keppler, G. Angres, H. R. Scherf, F. Wingen, F. Amelung and D. Schmahl, *Cancer Treat. Rev.*, 1990, **17**, 253.

228. T. Klenner, F. Wingen, B. K. Keppler, B. Krempien and D. Schmahl, *J. Cancer Res. Clin. Onc.*, 1990, **116**, 341.

229. T. Klenner, P. Valenzuela-Paz, F. Amelung, H. Muench, H. Zahn, B. K. Keppler and H. Blum, *Met. Complexes Cancer Chemother.*, 1993, 95.
230. S. Singh and S. S. Ray, *J. Nanosci. Nanotechnol.*, 2007, **7**, 2596.
231. Y. Wang, X. Wang, K. Wei, N. Zhao and S. Zhang, *J. Chen. Mater. Lett.*, 2007, **61**, 1017.
232. J. M. Xue and M. Shi, *J. Control. Rel.*, 2004, **98**, 209.
233. S. Prior, C. Gamazo, J. M. Irache, H. P. Merkle and B. Gander, *Int. J. Pharm.*, 2000, **196**, 115.
234. T. Y. Liu, S. Y. Chen, S. C. Chen and D. M. Liu, *J. Nanosci. Nanotechnol.*, 2006, **6**, 2929.
235. T. Y. Liu, S. Y. Chen, J. H. Li and D. M. Liu, *J. Control. Rel.*, 2006, **112**, 88.
236. J. A. Spadaro, T. J. Berger, S. D. Barranco and S. E. Chapin, *R.O. Antimicrob Agents Chemother*, 1974, **6**, 637.
237. K. Zhao, Q. Feng and G. Chen, *Tsinghua Sci Technol*, 1999, **4**, 1570.
238. L. Badrour, A. Sadel, M. Zahir, L. Kimakh and A. E. Hajbi, *Ann. Chim. Sci. Mater*, 1998, **23**, 61.
239. N. Rameshbabu, T. S. S. Kumar, T. G. Prabhakar, K. V. G. K. Murty and K. P. Rao, *J. Biomed. Mater. Res.*, 2007, **80A**, 581.
240. O. Palchik, J. Zhu and A. Gedanken, *J Mater Chem.*, 2000, **10**, 1251.
241. B. L. Cushing, V. L. Kolesnichenko and C. J. O'Connor, *Chem Rev.*, 2004, **104**, 3893.
242. T. Niidome and L. Huang, *Gene Ther.*, 2002, **9**, 1647.
243. H. Boulaiz, J. A. Marchal, J. Prados, C. Melguizo and A. Arenaga, *Cell. Mol. Biol.*, 2005, **51**, 3.
244. E. Orrantia and L. C. Chan, *Exp. Cell. Res.*, 1990, **190**, 170.
245. C. W. Pouton, K. M. Wagstaff, D. M. Roth, G. W. Moseley and D. A. Jans, *Adv. Drug Deliv. Rev.*, 2007, **59**, 698.
246. D. Luo and W. M. Saltzman, *Nat. Biotechnol.*, 2000, **18**, 33.
247. C. M. Wiethoff and C. R. Middaugh, *J. Pharm. Sci.*, 2003, **92**, 203.
248. K. M. Wagstaff and D. A. Jans, *Biochem J.*, 2007, **406**, 185.
249. S. P. Wilson, F. Liu, R. E. Wilson and P. R. Housley, *Anal Biochem*, 1995, **226**, 212.
250. F. L. Graham, A.J. van der Eb, *Virology* 973, **52**, 456.
251. P. Batard, M. Jordan and F. Wurm, *Gene*, 2001, **270**, 61.
252. I. Roy, S. Mitra, A. Maitra and S. Mozumdar, *Int. J. Pharm.*, 2003, **250**, 25.
253. S. H. Zhu, B. Y. Huang, K. C. Zhou, S. P. Huang, F. Liu, Y. M. Li, Z. G. Xue and Z. G. Long, *J. Nanopart Res.*, 2004, **6**, 307.
254. G. Bhakta, R. Singh, S. Mitra, S. Mozumdar and A. N. Maitra, *Proc. Control. Rel. Soc.*, 2003, **30**, 669.
255. A. N. Maitra, S. Mozumdar, S. Mitra, I. Roy, US Patent no. 6555376, 29 April, 2003.
256. S. Bisht, G. Bhakta, S. Mitra and A. Maitra, *Int. J. Pharm.*, 2005, **288**, 157.
257. D. Olton, J. Li, M. E. Wilson, T. Rogers, J. Close, L. Huang, P. N. Kumta and C. Sfeir, *Biomaterials*, 2007, **28**, 1267.

Subject Index